BRARY

~RAWN

Steps towards Life

Steps towards Life

A Perspective on Evolution

by

Manfred Eigen

with

Ruthild Winkler-Oswatitsch

Translation by

Paul Woolley

OXFORD NEW YORK TOKYO

OXFORD UNIVERSITY PRESS

1992

U2

Oxford University Press, Walton Street, Oxford OX2 6DP
Oxford New York Toronto
Delhi Bombay Calcutta Madras Karachi
Petaling Jaya Singapore Hong Kong Tokyo
Nairobi Dar es Salaam Cape Town
Melbourne Auckland
and associated companies in
Berlin Ibadan

Oxford is a trade mark of Oxford University Press

Published in the United States
by Oxford University Press, New York

Translated from the German edition, originally
published by R. Piper GmbH & Co, München, 1987

© Manfred Eigen, Ruthild Winkler-Oswatitsch, Paul Woolley,
and Oxford University Press 1992

All rights reserved. No part of this publication may be reproduced, stored in a retrieval system, or translated, in any form or by any means, electronic, mechanical, photocopying, recording, or otherwise, without the prior permission of Oxford University Press

This book is sold subject to the condition that it shall not, by way of trade or otherwise, be lent, re-sold, hired out, or otherwise circulated without the publisher's prior consent in any form of binding or cover other than that in which it is published and without a similar condition including this condition being imposed on the subsequent purchaser

A catalogue record for this book is available from the British Library

Library of Congress Cataloging in Publication Data
Eigen, M. (Manfred), 1927-
[Stufen zum Leben. English]
Steps towards life / Manfred Eigen: translation by Paul Woolley.
Translation of: Stufen zum Leben.
Includes bibliographical references and index.
1. Molecular evolution. 2. Life—Origin. 3. Molecular biology.
I. Title
QH325.E5513 1991 577—dc20 91–22349
ISBN 0–19–854751–X

Set by Pentacor PLC, High Wycombe, Bucks
Printed in Hong Kong

QH Eigen, M.
325
.E5513 Steps towards life
1992

577 Ei41s

Preface

The title of this book can be taken in two ways. First, the steps alluded to might be those first steps that evolution took — or ascended — towards the lowest level of life. For biologists, this first level is the cell, the smallest unit of autonomous life, and thus a forerunner of the single-celled organisms alive today. Fossils have revealed that this first stage of life had long been passed three thousand million years ago. The pre-cellular phase, which cannot have taken longer than the first thousand million years of our planet's existence, was astoundingly rich in invention and innovation. The most recent thousand million years have been no less extravagant: during this time, Nature has poured over the Earth a seemingly infinite wealth of species out of the cornucopia of evolution. So the fact that evolution is continuous in no way implies that it proceeds at an unchanging rate. Changes are prepared gradually, and then, suddenly, they break through and raise development to a new level. The transformation occurs sometimes in small steps, and sometimes in jumps which express a successful adaptation and often a completely new principle of operation.

This leads us on to the second possible interpretation of our title: steps which we ourselves take towards an understanding of the processes of life. Our insight also develops in steps on the large and on the small scale. This aspect is in fact the main aim of this book, that is, to make the principles of evolution clear and comprehensible, and to incorporate them into a unified physical world-view.

Molecular biology, which arose in the middle of this century from the disciplines of biochemistry and molecular structure determination, has gathered a momentum undreamed of at the outset of its short history. It is perfectly appropriate to speak of 'the era of molecular biology'. There is no shortage of excellent descriptions of this modern subject, with all its discoveries and the insight it has gained into structures and reaction mechanisms in biology. The only thing lacking in this new knowledge is its integration into a general understanding of Nature.

So far, such an attempt has been undertaken only once, by Jacques Monod. This was a fascinating and ambitious attempt, in which Monod did not shrink from drawing philosophical conclusions. It culminated in an apotheosis of chance. According to Monod, life can only be understood existentially. It can of course be reconciled with the laws of Nature, but it cannot be deduced from them. It is a pure creation from the nothingness of chance, not the revelation of a plan embodied in natural law. If it really were to emerge that there is only 'pure chance, absolutely free but blind, at the very root of the stupendous edifice of evolution', then this book would be superfluous. Our only task would be to report bald facts, dates, structures, and mechanisms. This would relegate biology to an existential enclave in the world-edifice of physics.

This book takes up the theme of Monod, whose plain language put many issues into clear perspective. But we shall not persist in proclaiming the omnipotence of chance, which has ruled over physics on the microscopic level ever since Maxwell and Boltzmann.

In his inaugural lecture at the University of Zurich in December 1922, Erwin Schrödinger declared: 'Physical research has shown clearly and unambiguously that for at least the vast majority of physical processes, whose regularity and reproducibility have led to the postulate of general causality, the common root of their strict, law-like behaviour — is *chance*.' These were the years before the Uncertainty Principle of quantum mechanics established chance as one of the foundations of physics. In biology, chance is reflected even at the macroscopic level: 'selection' implies that single, elementary events, determined by chance, are amplified autocatalytically up to visible numbers. None the less, law-like principles are also at work, and these are reflected just as much in the phenomena regarded as typically biological as in those associated with classical physics.

The arguments to be put forward here are based upon exact mathematical models and upon experimental studies of biological material. This book is intended to communicate new discoveries. The reason for its being written is similar to that for the writing of Charles Darwin's *The Origin of Species*. Darwin's view is accepted, just as the role of chance is accepted. However, this role will be interpreted in a way quite different from that current in biology.

The starting point for our discussion will be the epoch-making discovery made in 1953 by Francis H. C. Crick and James D. Watson, which ushered in the era of molecular biology. This was not so much the first description of the structure of deoxyribonucleic acid (DNA), based upon X-ray analysis, as the recognition that DNA is *the* molecule of heredity and that its structure holds the key to the understanding of heredity's molecular mechanism. The long-sought-after transition from chemistry to biology had been found. DNA is in itself a chemical substance, yet it is more than just a large molecule. By virtue of its chemical nature, DNA is an information store. This property, which goes beyond mere chemistry, is the determining force for everything else in biology. We shall come to discuss this in detail, even though this book is not intended as an introduction to molecular biology. Neither is it intended to describe the whole of evolution. Our performance will show just one act of this grand spectacle, the act spanning the period from the first nucleic acid molecules to the first cell, the period during which the transformation of inanimate to living matter took place. Ten chapters are dedicated to this 'day of creation', a day that lasted some five hundred million years. Each chapter is preceded by a quotation from the novel *The Magic Mountain*, by Thomas Mann.* *The Magic Mountain* appeared in 1924, when molecular biology was unheard of. So what is the relevance of these quotations?

This collection, selected and compressed, will convey the impression — perhaps more strongly than does the chapter 'Research' in *The Magic Mountain* — that Thomas Mann clearly occupied himself in great depth with the question that makes up the central theme of this book. It is that of the transformation of 'that nature, which did not even deserve to be called dead, because it was inorganic' into the 'simplest living organism'. And the reader will notice that Thomas Mann's reflections about life represent more than an aesthetic, literary counterpoint to the tenor of this book.

In the person of Hans Castorp, who dissects the living organism into smaller and smaller parts, Mann searches for the smallest living entities below the level of the living cell. 'Those were the genes'. But, he asks, can their 'elementary nature be established'? What do they look like 'after yet more light on the subject [is] forthcoming'? Mann reaches the conclusion that genes cannot be elementary structures in the chemical sense, but must in turn themselves have been assembled. In the manner of Hegelian dialectic, he sets up a contradiction, with the thesis that the elementary particles of life, genes, 'if they determine the order of life, ... must be organized' set against the antithesis that 'if they were organized, then they could not be elementary, since life depended upon organization'. He proceeds to resolve the contradiction with the synthesis: 'however impossibly small they were, they must themselves be built up, organically built up, with the order of life'. He even says what they were made of: 'molecular groups, which represented the transition between vitalized organization and mere chemistry'.

All this Thomas Mann wrote, as his diaries show, in 1920.† Since 1953, we have known how genes are built up, and how, within them, the transition from inanimate matter to the 'blueprint of life' takes place. The sub-units of genes are 'elementary' molecular groups in the chemical sense, chemical units. Only when they are linked up in the DNA molecule does a new, life-specific quality arise: *information*. Indeed, DNA is equipped with the most conspicuous properties of life. It has a memory, it can reproduce itself, it can mutate during reproduction and thus adapt itself by evolution, and by virtue of the metabolism of the cell it is prevented from sinking into a state of chemical equilibrium, which would exclude the possiblity of life. This little unit of life, says Mann, 'far below microscopic size' can grow spontaneously 'according to the law that

* Thomas Mann, *Der Zauberberg*, Fischer, Frankfurt, 1924; English translation by H.T. Lowe-Porter, *The Magic Mountain*, (Penguin Modern Classics), 1928. Quotations from *Der Zauberberg* in this book generally, but not always, follow Lowe-Porter's translation.

† Thomas Mann, *Tagebücher 1918–1921* edited by Peter de Mendelssohn, Fischer, Frankfurt, 1979.

each could bring forth only after its kind', and it possessed 'the property of assimilation' (adaptation) — all of these 'characteristics of life'.

Any of these quotations would befit the title-page of a modern textbook of molecular biology.

Where did Thomas Mann receive these insights that were thirty years ahead of their time? There are two indications to be found in his diary. An entry made on 14 June 1920 reads, 'Hertwig's *General Biology* arrived', and on 8 August, 'Reading Hertwig's *General Biology.*' Now and again, reading a borrowed textbook of physiology, he complains about 'the helplessness of science confronted with the actual process of life'. Apparently he found what he was looking for in the anxiously awaited work of Oscar Hertwig. Hertwig had made considerable contributions to biology both in teaching and in research. He had been the first to study the fertilization of the sea-urchin egg (1875), to observe the reduction in chromosome number in meiosis (1890), and to recognize the cell nucleus as the carrier of heredity. Although nucleic acid — the acidic substance found in the nucleus of the cell — had already been discovered by Friedrich Mieschner in 1869, it was not until 1943 that the role of nucleic acid as the material of heredity was demonstrated, by Oswald Theodore Avery. Oscar Hertwig was one of those whose ideas prepared the way for this discovery. In his textbook, the nucleic acids are mentioned explicitly as the presumed carriers of genetic memory. As a pupil of Haeckel, Hertwig was associated with the ideas of Darwin, although he later distanced himself from them. His *General Biology* contains all the technical terms and expressions that Thomas Mann later made use of, including the misinterpretations to which Hertwig had fallen victim, such as his allusion — in contrast to the beliefs of Darwin and Haeckel — to the inheritance of *acquired* characteristics. It comes as a surprise that Hertwig expresses himself in very much less pithy terms than does Mann, who obviously was endowed not only with the verbal ability to formulate the material but also with the keenness of thought to arrange it for himself: Hertwig discourses around all the ifs and buts of comtemporary critical opinion, which Mann simply omits. The mental processing of this complicated material took him less than three months, and on 30 September he records in his diary, 'Finished the biological chapter yesterday'.

Let us return to the present book. It is divided into three more or less independent sections. Part I is based upon an article entitled *Stufen zum Leben*, which appeared early in 1986 in a *Festschrift.* It was edited by Ruthild Winkler-Oswatitsch, who also provided the illustrations. These called for a supplementary text that was comprehensible in itself, and in this way the fifteen vignettes that make up the Part II of this book came about. These take up important individual problems and treat them independently. Thus the vignettes can be read or consulted without reference to the first part of the book. Although they are equivalent to the text in their general scope, they expand the text, make some points easier to understand, and go into somewhat greater depth. Whether the vignettes are read separately or just used as explanatory notes for the main text is a matter for the reader's own taste. Part III is a résumé that turned into an essay, and — as in a detective novel — this last part should perhaps be read first. In addition, the third part contains an epilogue, notes and literature references, and a glossary intended to assist readers who are less than familiar with molecular-biological terminology. The notes, too, can be read on their own. They contain sketches of the history of molecular biology.

This book would never have been completed without the tireless and — from frequent co-authorship — well-proven assistance of Ruthild Winkler-Oswatitsch, who took in hand the shaping of the text. The glossary and the notes are to a large extent her work. For this inestimable help she deserves especial thanks here.

Finally, I should like to acknowledge Paul Woolley's expert work in the preparation of the English translation of this book.

Manfred Eigen
Göttingen, July 1987 & November 1990

Contents

Part I: 'What was life?'-Theme and variations

Part II: Vignettes from molecular biology

Part III

Part I

'What was life?' – Theme and variations

1. Life is historical reality

What was life? No one knew. It was undoubtedly aware of itself, so soon as it was life; but it did not know what it was.

In any enquiry about the origin of life, we must make a clear distinction between the historical events that actually occurred and the conceivable events that natural laws make possible in principle. The study-matter of biology is the world of living organisms, in which the historical process is made manifest. The reconstruction of the course of evolution is restricted to using historical evidence. Such evidence as we possess indicates that all forms of life have a common origin. At every level of life, we find not only individual variability but also similarity of detail and the universal validity of the underlying physical and chemical laws. We can recognize these laws only if we abstract from reality; for physics is concerned not with individual processes as such, but solely with the repeating regularities in the processes.

Take an example from meteorology. On a sultry summer's day we see a storm coming towards us, and can make out all kinds of strange shapes and patterns in its towering cloud formations. The physical effect that causes clouds — the condensation of water vapour caused by changes in pressure and temperature — is from the physicist's point of view almost completely understood. An 'all-or-nothing' transition from a gaseous to a liquid state gives rise to the material of the clouds, and currents of rising warm air shape the formations that we see. Yet no physicist could predict any of these structures quantitatively, or in detail.

Life is *not* an inherent property of matter. Life is indeed associated with matter, but it appears only under very specific conditions and, when it does, it expresses itself in very diverse and individual ways. It is therefore perfectly logical to set the question of life's origin alongside that of its nature. We shall come closest to understanding the principle of life if we can discover the principles according to which life *could* begin. This is a challenge addressed to the physicist, even if he calls himself a biophysicist, a biochemist, or a molecular biologist. How life *did* begin, however, can probably only be understood by appeal to historical evidence. Although the purely historical question will be of less interest to us here, we shall start by presenting it, in order to develop a more precise idea of what life is.

Many misunderstandings have resulted from ignoring the difference between these two questions. There is no general physical theory that explains the *historical* origin of life. How life commenced must be regarded as a succession of events whose details can be neither reconstructed nor predicted. But they took place, none the less, under the directing influence of natural law.

2. Can the historical origin of life be reconstructed?

What was life? No one knew. No one knew the actual point from which it sprang, where it kindled itself. Nothing in the domain of life seemed uncaused, or insufficiently caused, from that point on.

Can we reconstruct the events that led to life? The answer depends upon how much evidence of their historical course we can find. The most recent phase of evolution can be charted with the help of fossils whose age is determined by the geological dating of their excavation sites. In recent decades, the use of radioactive isotopes has made dating more accurate, and has enabled it to reach further back into the past.

In order to assign species to their correct places on their family tree, we need to compare their characteristics with as much care as possible. However, our knowledge of these becomes increasingly sparse and sketchy when we go back beyond the Cambrian period, which began over 600 million years ago. For multicellular organisms, which originated at the beginning of the Cambrian period, it has proved possible to draw up a phylogenetic family tree.[1] However, although they dominate life today, multicellular organisms are relative newcomers; single-celled organisms populated the Earth at least three and possibly four thousand million years ago.[2] Unfortunately for palaeontologists, single-celled organisms (at least fossilized ones) look largely alike. This makes it an impossible task to compare their characteristics with an exactitude sufficient to determine precise phylogenetic relationships.

Our actual subject is not the development of species but the reconstruction, as far as possible, of their historical *origin*. For this purpose, we need to refer to the earliest branching-points of the phylogenetic family tree.

Can we regard the existence of such a tree, with successive branching-points, as an established fact at all? It is especially important to have a clear answer to this, as the objection is often raised that the construction of phylogenetic trees involves the assumption of successive branching. If this were true, the conclusion that a family tree exists would be based upon a circular argument.

The assumption of successive branching in phylogeny is indeed inescapable, if we start off by assuming that species really did develop by evolution and can be traced back to a single ancestor. But in order to *prove* evolutionary development, we would need first to *deduce* a pattern of consecutive branching, as an empirical result from available data. If a branching pattern emerged, and turned out to be clearly distinguishable from alternative patterns — such as bundle-like or criss-cross ones — then we could take it as strong evidence for the historical reality of evolutionary development. We therefore seek a procedure to help us deduce precise topologies of branching.

There is one method that offers a quantitative solution for these questions, and it has been developed over the last two decades to a point where it can be applied seriously to them. The method is the comparative analysis of protein and nucleic acid sequences.

Since the 1950s, it has been known to biologists that the blueprints for the construction of organisms (and therefore for all the molecular functional units that living cells construct and use) are encoded in a linear form, like written documents, and are laid down in giant chain-like molecules of deoxyribonucleic acid, or DNA. Individual portions of DNA that encode a single molecular functional unit, a protein molecule, are called genes. The 'genetic information' that they contain is the sequence of its basic building-blocks, or symbols, chemically termed *nucleotides*. These basic units in a stretch of nucleic acid are translated, one after another, into a corresponding linear sequence of protein units. Determining the sequence of nucleotides enables one to identify the genetic information.

This is not the place to describe the rather complex biochemical procedures used for the determination of the sequences of DNA and proteins. These methods are today — especially for DNA — a matter of routine. Used together with the new gene technology, which allows the DNA of a gene to be divided accurately into sections, these methods could in principle be used to decode and to document all the gene sequences of all organisms. Today we are witnessing the emergence of a completely new kind of library, genetic libraries, whose documents are stored in an international network of data banks that can be accessed from suitably equipped laboratories anywhere on Earth.[3]

What could be more logical than to draw upon these DNA sequences in order to carry out a quantitative comparison of degrees of kinship between different organisms? To do this, we simply take a given gene and compare its different sequences, symbol by symbol, in the organisms chosen for comparison. The kinship between the organisms must naturally be strong enough for the common overall pattern to be visible. Each pair of sequences from the group is then characterized by what we may call an *information distance*. This is equal to the number of positions in the sequences which would be expected to correspond to one another, but which turn out to be occupied by different symbols. The complete set of such distances, for all possible sequence pairs, is then called a *distance space*, and its branching pattern can be determined objectively by a mathematical procedure.[4] It is worth mentioning that the mathematical treatment goes far beyond the mere determination of distances between pairs of sequences: it also takes account of the position-specific distribution of homology among combinations of sequences (Vignette 1).

In order to determine the divergence between sequences, we distinguish three basic kinds of topology. A *bundle-like* branching pattern can be expected when the related sequences possess a single precursor, and its descendants have developed independently of one another and diverged by simultaneously accumulating mutations over many generations. A *tree-like* topology, in contrast, is the result of successive divergences: a sequence splits into two daughter sequences that mutate and branch independently over a long period. The third variant is a criss-cross network, typical of colonies of mutants. This kind of network of mutants is found with sequences that arise perpetually from a single, optimized sequence, because identical mutants can arise independently by different paths, especially if they are closely related to the optimized sequence. This is always the case when the total number of symbols in a sequence is so small, and the time interval considered is so great, that every symbol gets several chances to mutate. An initially bundle-like or tree-like distribution of mutants ultimately turns into a network-like distribution, through the repeated appearance of cyclically closed patterns. Here we borrow an expression from communications theory, and speak of a randomization, or blurring, of the topology by noise. In the limit of extreme randomization, it is no longer possible to extract historical information from the sequences.

For a successful application of comparative sequence analysis, it is therefore necessary that

- the sequences be sufficiently long,
- the distances between the individual sequences — as compared with the sequence lengths — be neither too great nor too small, and
- a sufficiently large number of different, related sequences be available for analysis.

The first two conditions arise in consequence of the random way in which mutations occur. To be of any value, the statistical significance of the result must be sufficiently high. In order for this to be achieved, the distances between sequences must be well below the length of the sequences. For example, any kinship between binary sequences (consisting of two symbol types only), separated by a distance equal to half their length, would not be detectable, since the probability of two symbols being the same will be 50 per cent on a basis of chance alone. If the sequences are short and the divergence between them is considerable, then many 'undetectable' mutations will be present; these can be due either to parallel mutation, in which two sequences have mutated independently at the same position in a similar way, or to back mutation, in which a mutation in one generation is reversed in a later generation. Such undetected mutations can be estimated and corrected for in the mathematical treatment, as long as they are not too numerous.

The third and final condition is an important prerequisite for determining as accurately as possible the topology of the sequence space under analysis. Today we can register only those particular mutations that have been preserved by evolution. In this preservation, both the capacity for change (the error rate in reproduction) and the stabilization of change play a part. In other words, it is not enough that mutants appear from time to time; they must also spread through the population and come to dominate it. However, the appearance and the stabilization of mutants must remain within certain limits, so that the original information is not destroyed. If that took place, then any similarity remaining between two sequences would be purely coincidental.

Fortunately, Nature has a large repertoire of gene sequences, with greatly differing degrees of phylogenetic divergence (Vignette 2). We find, on the one hand, highly variable genes, by the help of which even the degrees of kinship between man and non-human primates can be gauged, and, on the other, genes which are so highly conserved that they can lead us back to primordial branching-points such as those within single-celled organisms like eubacteria or archaebacteria. There are even gene families whose kinship goes back to the pre-cellular stage. These can give us, for example, a glimpse into the origin of the genetic code.

We are thus in a position to deduce a whole row of conclusions from the investigation of sequences, and to answer the question with which we introduced this chapter.

All stages of evolution, from the differentiation of the primates back to the very first branching of single-celled organisms, can be analysed quantitatively by the comparison of suitable genes or proteins. Of the very many protein sequences determined, the families best investigated include at present the immunoglobulins, the red blood protein haemoglobin, and the respiratory enzyme cytochrome c. For the nucleic acids, it is above all the components of the cell's apparatus of reproduction and translation whose sequences give an eye-witness account of the first stages of biological evolution. These have already been analysed for hundreds of species, and they show clearly the earliest branching-points of evolution.

The topology of all *phylogenetic* kinship is unambiguously tree-like. Deviations are so small that this branching topology can be distinguished clearly from all others.

In contrast to the successive, tree-like divergence found in phylogeny, other sequence families, such as the various tRNA molecules within one and the same organism, spread out in a bundle-like way. These are clearly sequences that have arisen from a single precursor and then developed independently of one another.[5]

The sequences of 'proto-genes' — the very earliest genes — can to some extent be reconstructed. They show characteristic patterns indicative of a simple, primaeval code. They are particularly rich in the monomers G and C, which impart stability to the folded structures of nucleic acids. Phylogenetic analysis of the descendants of these molecules reveals the pattern of the primaeval code the more sharply, the further back in time one goes. This is an especially important finding, since the mere observation of similarity between two sequences does not necessarily point to a common ancestor; it can equally indicate an incomplete, or arrested, adaptation towards the same end. It allows us to conclude that the nucleotide patterns preserved down to the present day are documents of the primaeval era.

At present, the very first branching-points cannot be dated with exactitude. We know only the absolute distances between the sequences that have been analysed and the resulting upper and lower limits on a relative time-scale. In order to transform the information distances into an absolute time-scale, it would be necessary to know the relevant mutation rates, that is, the rates at which mutations become fixed. For the geologically accessible phase of evolution, into which most of the branching-points for eucaryotes fall, such a transformation can easily be carried out, allowing the branching to be dated approximately.

However, for the initial stages of evolution, we can do no better than place limits upon the time-scale. For the very first phylogenetic branching, this goes back about three thousand million years. The extrapolation to pre-cellular stages is uncertain, but the earliest branching-points of molecular functional units that can be reconstructed from sequence distances — for example, the adaptors of the genetic code — lie quite close to the branching-points for the first cells (Vignette 3). Conservative estimates place the phase of the origin of the cell's molecular tool-kit no more than about four thousand million years ago (the best present estimate is 3800 ± 600 million years). Thus, the origin of the genetic code and the cell's translation machinery, followed by their integration into a unit to be imposed on all later life forms, took less than a quarter of the entire period of evolution on Earth.

There is much evidence in support of the conclusion that life arose on our planet, the age of which is some 4700 million years. There is certainly no known reason to doubt that this age is sufficient to account for the self-organization of reproductive molecular systems.[6] The evolution of species, from the first unicellular organisms to humans, took place within three to three and a half thousand million years. It reveals itself in reconstructable phylogenetic trees. The prospect of documenting the entire genetic information of many organisms within the foreseeable future opens further possibilities for a well-founded reconstruction of the historical origin of life.

3. Complexity as a physical problem

What then was life? It was warmth, the warmth generated by a form-preserving insubstantiality, a fever of matter, which accompanied the process of ceaseless decay and repair of albumen molecules that were too impossibly complicated, too impossibly ingenious in structure.

If we regard the phenomenon 'life' as a regularity in the behaviour of matter, we must at some point ask the question: 'What kind of physical principle lies behind this behaviour, and what are its effects?'

The question 'What is life?' has many answers, none of which is ultimately satisfying. (Thomas Mann points up this enigma again and again, continually repeating the question and reshaping the answer.) The manifestations of life, and the characteristics and capabilities of living beings, are too numerous and too various to allow a meaningful general definition; such a definition would not be able so much as to hint at the individuality and the variety that make up the essence of life. The reason for this lies in the complexity common to all the forms of life with which we are acquainted. At the molecular level, the same problem is encountered in the structures associated with the processes of life: the nucleic acids and the proteins.

How complex are the most primitive organisms? Even this simple question seemed for many years to be unanswerable. Yet today we know that every organism is represented by a 'blueprint'. This is handed down from one generation to the next, and it ensures that each generation of progeny resembles its parents. While this applies especially to vegetative reproduction, it also applies with certain restrictions — those that are imposed by the nature of the genetic crossing process — to sexual reproduction. The problem of the complexity of organisms can thus be reduced to that of the complexity of their blueprints, since it is these that, in an appropriate environment, provide the instructions for the origin and the development of an organism.

Because the genetic information lies in the blueprint as a linear sequence of symbols, our question takes us ultimately to the quantity of information that can be put into a sentence — first of all, the absolute amount of information rather than its semantic content. There is a mathematical theory, information theory, that has a quantitative answer ready for us.[7] The information content is the average number of binary yes/no decisions that are necessary in order to identify unambiguously a particular sequence of symbols. If all possible arrangements of the symbols in a given sequence were equally probable, one would have to go through all the alternative symbol sequences in order to hit the right one by chance. In a case like this, the number of possible sequences that can be produced from a defined set of symbols gives a measure of the information content. The reciprocal of this number gives the probability of the appearance of a *particular* sequence. Since the numbers of sequences are additive, while probabilities are multiplicative, it is usual to state the quantity of information not by the number of alternative arrangements but by its logarithm, as the logarithm of a product is equal to the sum of the logarithms of the factors. Further, logarithms to base 2 are used, and the unit in which the result comes out is the binary digit (*bit*). The bit number corresponds to the length of a sequence of binary characters. The nucleic acids employ four symbols, so that a gene of length N has 4^N (or 2^{2N}) different sequences. On the assumption that these are all equally probable, this means that the information content of the gene amounts to $2N$ bits.

How complex are organisms? The smallest autonomous units, unicellular microorganisms such as the bacterium *Escherichia coli*, have incorporated into their genomes a few millions of symbols — roughly the equivalent of a thousand-page book. The number of symbols in the genome of the human being is nearly a thousand times larger. It represents a respectable library. It would be pointless to try to imagine the number of alternative arrangements of the letters; our imagination simply does not run to such numbers. Consider instead just a single gene, with only a thousand symbols: this is like a sentence in the genetic language, and corresponds to one functional instruction. With four classes of symbols — so that each of the thousand positions can be occupied in any of four different ways — the number of variants that result is $4^{1000} \approx 10^{602}$ (a one followed by 602 zeros). The volume of the entire universe, calculated as a sphere with a diameter of ten thousand million light-years, amounts to a 'mere' 10^{84} cubic centimetres, or 10^{108} cubic Ångström units. The entire material content of the cosmos corresponds, weight for weight, to fewer than 10^{75} genes of the length assumed in this example.

These numbers, related to the size of the universe, are in any case quite irrelevant for our discussion. They serve simply to demonstrate our complete inability to conceive in any realistic way of numbers like 10^{600}. It would be more interesting to find out how many molecules, of length one thousand symbols, could have been tried out in the course of the process of evolution within the spatial and temporal confines of our planet. Naturally, the process of evolution cannot be reconstructed in such detail as to allow a precise answer to this question, but the number we are looking for probably lies between 10^{40} and 10^{50}. If we covered the Earth with a layer of solution one centimetre thick, containing nucleic acid at a concentration of one gram per litre, and allowed the nucleic acid molecules in it to form and decay with a lifetime of not more than one second each, then after one thousand million years there would have arisen some 10^{50} fresh molecules.

It is possible to make a similar estimation on the scale of the laboratory. A research student synthesizing nucleic acid molecules enzymically in a one-litre flask could, by working for twelve hours each day for a whole year, produce 10^{25} sequences. Compared with the planetary scale, that does not sound entirely hopeless, especially as the conditions for natural synthesis may have been assessed rather too optimistically in many respects. Realistic concentrations of nucleic acids are probably orders of magnitude lower than originally assumed. Genes arise by reproduction; that is, if a gene sequence occurs at all, then it occurs with high redundancy. Furthermore, it is not true that the entire surface of the Earth was available for the reaction, or that the yield was as high as supposed. But there still remains a conflict between the orders of magnitude, the discrepancy between what was possible and what would be necessary if genes were the product of a purely random synthesis.

What conclusions can we draw from this?

The genes found today cannot have arisen randomly, as it were by the throw of a dice. There must exist a process of optimization that works towards functional efficiency. Even if there are several routes to optimal efficiency, mere trial and error cannot be one of them.

The discrepancy between the numbers of sequences testable in practice and imaginable in theory is so great that attempts at explanation by shifting the location of the origin of life from Earth to outer space do not offer an acceptable solution to the dilemma. The mass of the universe is 'only' 10^{29} times, and its volume 'only' 10^{57} times, that of the Earth.

The physical principle that we are looking for should be in a position to explain the complexity typical of the phenomena of life at the level of molecular structures and syntheses. It should show how such complex molecular arrangements are able to form reproducibly in Nature.

4. How does information arise?

Seeking a connecting link, they had condescended to the preposterous assumption of structureless living matter, unorganized organisms, which darted together of themselves in the albumen solution, like crystals in their mother-liquor; yet organic differentiation still remained at once condition and expression of all life. One could point to no form of life that did not owe its existence to procreation by parents.

We have already encountered the key-word that represents the phenomenon of complexity: information. Our task is to find an algorithm, a natural law that leads to the origin of information. The definition of information given on page 9, based solely on the number of possible arrangements, is incomplete. It applies only to the limiting case in which all arrangements are equally probable. If we try to guess a binary sequence by asking questions that can be answered only with a 'yes' or a 'no', then the average number of questions needed will correspond to the information content, or bit number, of the sequence, assuming that the two symbols have the same expectation values at each position. If the sequence is a sentence in a human language, then knowing the code that relates the letters to the binary digits (as in a teleprinter), we could arrive at our goal much more quickly by taking account of the familiar properties of the language in making our guesses. Analysis of the frequency of symbols in English shows that, apart from the space between words, the letter 'e' is the most common. The average length of words is governed by the use of the space symbol. In the construction of words, we know that, for example, 'q' is always followed by 'u'. We know as well that vowels and consonants are not distributed arbitrarily. The structure of sentences is largely determined by the usage and the order of articles, nouns, adjectives and verbs. There are also grammatical and syntactical rules. The spectrum of rules and conditions covers even the meaning of the sentence, that is, the semantic information, which is dependent upon specific premisses that cannot be described by general statistical frequency laws.

Claude Shannon, one of the founders of information theory, once got his students to play the following game. One player thinks of a sentence. The other must guess this letter by letter. His questions can be answered only with a 'yes' or a 'no'. The number of questions asked before the sentence is guessed is then compared with the number that would have been necessary without taking account of the known structure and rules of language. An analysis of one hundred examples revealed a reduction of the information needed from 4.76 bits per symbol for a fully uncorrelated distribution of letters to 1.4 – 1.94 bits per symbol for an English sentence. Naturally, the players use their entire knowledge of the structure of language in guessing the unknown sentence.

We conclude that every constraint that makes the distribution of prior probabilities of the symbols less even will reduce the quantity of information that is needed for their identification. This quantity corresponds to an average number of bits per symbol, multiplied by the total number of symbols in the message. It is not just a question of the absolute number of symbols and the total number of alternative arrangements, but also one of the average realizability of the alternatives. The origin of information is thus tantamount to a change in the probability distribution of the symbols on the basis of additional constraints that first emerge during the evolutionary process.

There exists a direct relationship between the quantity of information as considered above and the quantity known in thermodynamics as entropy: the information defined by Shannon corresponds to negative entropy. (These relationships are treated in textbooks and will not be explained further here.[8]) For our present discussion, it is of greater importance to recall that at thermodynamic equilibrium the entropy has reached a maximum. Any perturbation of the equilibrium produces a reaction that proceeds in a direction such as to cancel out the perturbation. Equilibrium is a stable state. There are no perturbations that can change the probability distribution in the system, as long as the external conditions are unaltered. Thus, information can *not* arise in systems that are in thermodynamic equilibrium.

So how has the information in genetic blueprints, the fixation of particular arrangements of symbols, come into being? A biologist would answer: by natural selection! He would add that the gene sequences contained in organisms, coding as they do for functions that are optimally adapted for life, are in fact the products of a whole series of changes in the sequence, stabilized one after another by selection. Such a process need by no means take place at a uniform rate. From time to time, stable intermediate states will be reached, during which evolution apparently comes to a standstill, because for a while no advantageous mutations occur. These are followed by phases of change, either due to rare mutations or caused by environmental factors that set off a cascade of successful mutations. In this way, intermediate states come about which are not yet optimally adapted and whose mutant spectra soon give rise to better-adapted sequences. This makes it appear as though Nature had been making jumps, because the relatively short-lived intermediate states leave no trace behind them.

Darwin's principle brings about what theoreticians would call the generation of information. Dominance by a wild type established by selection means the local stabilization of a particular probability distribution. The appearance of an advantageous mutant destabilizes the hitherto stable state, and establishes a new probability distribution, which in turn can be destabilized by another new mutant, and so on. The fact that a destabilizing mutant can appear at all shows that selection has nothing to do with genuine equilibrium states. It is true that statistical fluctuations (the general cause of mutations) do occur within systems in equilibrium. However, in equilibrium they cannot amplify themselves so as to become macroscopically observable. Every mutant — even an advantageous one — starts life as a single copy. Its ability to multiply in number depends upon special criteria that are not realized in equilibrium, and which are based solely on the properties of the constituents of living systems.

In order to illustrate the law-like character of the selection process, we consider the law of mass action, which governs *chemical* equilibrium. In a closed chemical system, which does not exchange matter with its surroundings, the proportions of all components inevitably and reproducibly reach those values dictated by the law of mass action. Such an equilibrium distribution can be regarded as 'selective'. The basis of the selection is the free energy (that is, energy available for doing external work) contained by the individual components. Components with a lower free energy are preferred. However, this does not mean that other components, with higher free energy, die out. They simply appear less frequently, in accordance with the balance that follows from the law of mass action. As long as an equilibrium between two components can be defined, both will appear with finite frequency.

In the case of the Darwinian selection principle, the problem raised is very similar, but the answer is quite different. We look for the selection of a particular genotype, the one that encodes the phenotype with the quality of being 'best adapted', meaning 'producing progeny at a maximum rate'. The genotype appears first as a copy of a sequence already in existence. But there always occur errors in copying. An error can be a symbol incorrectly copied, or the omission or addition of single symbols or even of entire blocks of symbols. Such mutations are the source of evolutionary progress. Selection now has two aspects. First, there is a single sequence that among all the mutants corresponds to the best-adapted phenotype (the 'wild type'), and its growth outstrips that of the rest of the population. The less well adapted mutants — according to the classical interpretation — die out. They cannot coexist with the wild type and can at best appear sporadically as statistical fluctuations. The selected sequence becomes the basis of further progress. Secondly, the information in the wild type is conserved as long as there is no better-adapted variant among the mutants that have been randomly 'tried out'. In this way errors are prevented from accumulating. This requires that the error rate remain below a certain, critical threshold level. Only when a variant appears which is (even) better adapted does this take over the role of the wild type.

Returning to the question with which this chapter is headed, we could perhaps regard natural selection as the key to the problem of information and complexity. But what mechanism guarantees selection and, with it, the inevitable appearance of information?

5. Life is a dynamic state of matter organized by information

Those were the genes, the living germs, bioplasts, biophores .... If they were living, they must be organic, since life depended upon organization. But if they were organized, then they could not be elementary, since an organism is not single but multiple. They were units within the organic unit of the cell they built up. But if this was true, then, however impossibly small they were, they must themselves be built up, organically built up, by a law of their existence.

Information is stored in DNA, the blueprint of organisms. This sequence of symbols must be organized, as in a language. Indeed, there is a form of punctuation, or subdivision of contents, that divides up the enormous document into words (codons), sentences (genes), paragraphs (operons), and entire volumes (chromosomes). This organization is genetically fixed; it is laid down in the structure, that is, in the sequence of the chemical units (nucleotides) of the DNA molecule.

But how has the information-rich, ordered state of this molecule come about? In terms of structural stability, a molecule carrying useful information has not the slightest advantage over a molecule carrying nonsense. The strength of the chemical interactions that stabilize the molecule and preserve the information contained in the symbol sequence, so that it can be handed down from one generation to the next, does indeed depend in part upon the composition of the sequence, but it has nothing to do with the information stored within it. The structural stability of the molecule has no bearing upon the semantic information which it carries, and which is not expressed until the product of translation appears. The selection of 'informed' molecules is not based upon structural stability, but upon a kind of order that lies in the selection dynamics of its reproduction.

Consider, for instance, one of Mozart's compositions, one that is retained stably in our concert repertoire. The reason for its retention is not that the notes of this work are printed in a particularly durable ink. The persistence with which a Mozart symphony reappears in our concert programmes is solely a consequence of its high selection value. In order for this to retain its effect, the work must be played again and again, the public must take notice of it, and it must be continually re-evaluated in competition with other compositions. Stability of genetic information has similar causes.

Before we look into the question of the origin and the generation of the dynamic order of 'life', we must ascertain whether this order really is of relevance for our problem.

What do all living beings have in common?

They all use DNA as a store for their hereditary material, processing the stored information according to the scheme:

Legislative	→	Message	→	Executive	→	Function
DNA	→	RNA	→	Protein	→	Metabolism

Not only is this general scheme common to all organisms on Earth, but so is its detailed structure too. All organisms make use of a common genetic code, a common biochemical machinery, and synthetic macromolecular products that are organized according to common structural principles (Vignettes 4 – 8).

The lowest unit of autonomous life is the cell, the prototype of which is also built up according to a common design, even though the cells of multicellular organisms differ greatly in their observable function. Whatever task a cell is adapted to, it carries out with optimal efficiency. The blue-green alga, a very early product of evolution, transforms light into chemical energy with an efficiency approaching perfection. The enzymic reactions in a bacterium hardly differ in their efficiency from those that proceed inside the human cell. Archaebacteria show their ability to survive under extreme environmental conditions such as high salt concentration or high temperature. The amoeba is also a single-celled organism, but this time at a much higher level of development; it displays a form of social behaviour, by communicating chemically with its fellows, and by cooperating in sporulation for the production of the next generation. The higher organisms demonstrate the most marvellous and astounding achievements. From loosely associated agglomerates of cells there have emerged centrally directed, multicellular forms of organization with highly differentiated functional demarcation.

All the varieties of life have a common origin. This origin is the information that, in all living beings, is organized according to the same principle.

An understanding of this principle will bring us closer to an answer to the question asked at the beginning of this book: 'How *can* life begin?'. Just as life has passed through many stages of development, there must also be many principles of organization: for the reproduction of individual genes, for their cooperative integration into a functional unit, for regulated growth, for the construction of cellular structures, for recombinative inheritance, for the differentiation of cells, and for the construction of organs up to and including the construction of *the* organ that functions as a memory, the organ that itself stores, processes, and creates information, bringing about a new kind of evolution at a level above that of matter.

6. Is there a principle of order in biological systems?

But what was all this ignorance, compared with our utter helplessness in the presence of such a phenomenon as memory, or of that other more prolonged and astounding memory called the inheritance of acquired characteristics?

The principle of order upon which we shall now focus our attention is intended to explain how information comes into being. It will have to be a dynamic principle. Information *arises* from non-information. We are not merely dealing with a transformation that makes existing information visible. The state of the system has a completely new quality after information has arisen. The new information has made the previous, information-deficient state unstable, and thus has consigned it irretrievably to the past.

That is how a physical interpretation of the Darwinian principle might sound. According to Darwin's principle, whatever is better adapted spreads out and displaces its less well adapted predecessor. Thus, complexity, built upon simplicity, has accumulated throughout biological evolution from the first single-celled organisms to human beings. Evolution as a whole is the steady generation of information — information that is written down in the genes of living organisms.

Reading Darwin,[9] one becomes aware that he intended his principle to be applied to biological reality; he did not regard it as an abstract principle of order, but as a tendency directly visible in Nature, even though it was accompanied by many ifs and buts that later were to be objects of research by population biologists. Only once, in a letter to Nathaniel Wallich (1881), did Darwin reveal his presumption of a general principle of order behind this process of life. In this letter, he asserted yet again that the question of the origin of life was, with the state of knowledge of his time, unanswerable, *ultra vires*, and he emphasized that his ideas referred only to phylogenetic descendence. Yet then he added: 'I believe ... that the principle of continuity renders it probable that the principle of life will hereafter be shown to be a part, or consequence, of some general law.'

Today we can apply our knowledge to molecular systems such as genes and the products of their translation. We can also investigate in a much more objective way the physical nature of the Darwinian principle: theoretically, by defining accurately the prerequisites and constraints, and experimentally, by exact control of experimental conditions.[10] We find that the selection principle is neither a mystical axiom immanent in living matter nor a general tendency observable primarily in living processes. On the contrary, it is — like many of the known physical laws — a clear 'if-then' principle, that is, a principle according to which defined initial situations lead to deducible behaviour patterns. It is thus analogous to the law of mass action, which regulates the attainment of the quantities of the components in a chemical equilibrium.

The initial situation must fulfil the following prerequisites.

- The individuals (DNA molecules, viruses, bacteria) among which selection is to take place must be self-reproducing. Once they have come into existence, they can multiply by the copying of individuals already present, but not by synthesis *de novo* of new individuals. We shall call these self-copying individuals *replicators*.
- The first condition just stated is modified by allowing the self-reproduction to be subject to error. This is ultimately because the physical process of copying takes place at a finite temperature, and the energy associated with the interactions involved in copying is of the same order of magnitude as thermal energy. The molecules involved in replication are thus subject to the buffeting of thermal motion, which results in mutations. This means that some replicators come into existence not as the result of true copying of an identical parent, but in consequence of inaccurate copying of one that is closely related.
- The self-replication must take place far away from chemical equilibrium. This means that the system of replicators requires a perpetual supply of chemical energy. In other words, the system must possess a metabolism. In chemical equilibrium, by contrast, formation and decay at the chemical level are strictly reversible: the autocatalytic formation caused by self-replication would compensate the autocatalytic decay and would therefore be unable to effect selection. Information cannot originate in a system that is at equilibrium.

Self-reproduction, mutagenicity, and metabolism are necessary conditions for natural selection. In a system that possesses these three properties, selection automatically goes into action. The consequences are seen most clearly in terms of the relative population numbers: one of the many replicators increases in number until it comes to dominate the entire population. Even for the smallest differences in selection value, the result is an all-or-nothing decision, as long as the competing replicators are unrelated. However, mutants closely related to the dominating replicator will be tolerated, according to their own selection value and the closeness of their kinship. This implies that replicators of (nearly) equal selection value, that are closely related, will share the dominance. This is true irrespective of whether the system is stationary, is growing, or varies in any other way with time.

The selection value is a parameter defined by the dynamic properties of the system, and it takes account of the rate and quality of reproduction and the lifetime of a replicator. If selection values depend upon environmental conditions, such as the presence of accelerating or interfering substances, then their mathematical expression can be quite complicated. However, for simple molecular systems under defined conditions, selection values can be measured by the methods of chemical kinetics.

For large population numbers, behaviour in selection is as regular as the attainment of chemical equilibrium. It is also as inevitable, as long as the conditions noted above are fulfilled. Selection contains an element of exact 'if–then' behaviour. It has nothing to do with the tautological interpretation 'best adapted = selected'. 'Selection' could in principle just refer to *any* kind of preference. But here it means a *particular* kind of preference, which adheres unerringly to a single scale of values. Selection is based upon self-replication. It distinguishes sharply between competitors, it constructs a broad mutant spectrum on the basis of value, and in this way it organizes and steers the entire, complex system. It is true that the mass action of chemical equilibrium results in a kind of selection of the structurally stablest configuration. But this selection does not share the exclusive nature of the selection between replicators, which is inherently associated with non-equilibrium situations. High structural stability in chemical equilibrium is purchased at the price of extreme lethargy of the reaction. In contrast, the replicator that is stabilized dynamically in a non-equilibrium system dies out within a short time if a superior competitor appears.

Since all organisms are self-reproducing, selection plays a decisive role for all organisms. Naturally, there are variations associated with various internal and external secondary conditions. Under particular external conditions, selection can turn right round into coexistence or even cooperation. For example, if two or more different replicators are coupled in a cyclical way by repression or promotion effects, then the result is a regulated coexistence of all the partners, and the cycle as a whole competes with other such cycles. The higher organisms in particular have special mechanisms of inheritance: they exchange genes by recombination, for which the selective evaluation takes place at the level of the organism as a whole. It was certainly these organisms that Darwin had in mind when he formulated his selection principle. The abstract, mathematically deducible principle is associated with clearly defined conditions such as self-replication with mutation and non-equilibrium. These are fulfilled ideally by single DNA or RNA molecules, by genes, and likewise by viruses and vegetatively reproducing organisms. Self-replication can even be realized in non-living systems. Thus, laser modes reproduce themselves by resonance, and we discover selection, the dominance of a particular mode, in the light of a laser. The theory of this was developed by Hermann Haken[11] in parallel with the theory of molecular selection.[12] The mathematical structures of these two theories are strikingly similar.

The selection theory of self-reproducing systems represents something quite new in physics, as it establishes a value function defined in terms of dynamic parameters. The presumption that calling Darwin's principle 'the survival of the fittest' reveals its tautological nature is thus pursued *ad absurdum*. 'Survival' is an empirical fact that can be expressed in relative population numbers and measured experimentally. 'Fittest' is determined by a value function. It is based upon dynamic parameters that are measurable independently of the population numbers. For molecular systems, the connection between these two is accessible not only for exact formulation but also for quantitative measurement. In fact, it is the 'how' of selection, and is just as little a tautology as is a physical law such as Einstein's energy–mass relation.

Selection implies focusing upon one out of many alternative sequences. This dominant sequence, the so-called wild type, is the sequence most frequently found. The wild type is represented by more individuals than any other mutant. However, since there are many mutants, the wild type still only makes up a small percentage of the total number. If there occurs a change in the environment, then in general the rich spectrum of mutants can provide a better-adapted variant that will grow in number.

This fact came as a surprise to biologists. It had been supposed, because the wild-type DNA sequence can be determined unambiguously, that most of the individual sequences in the sample under investigation actually were that of the wild type. The implicit assumption was that only an absolutely dominant sequence could be determined exactly. However, it was a facile argument. The wild-type sequence can easily be present in a small number, as long as the many mutants are distributed around the dominant sequence. In this way, the average of all sequences is identical with the individual sequence of the wild type, even if the latter is present in vanishingly small amounts or not at all. Such a distribution we call a *quasi-species.*[13] It is the target of selection. The quasi-species, with all its mutant members, is selectively evaluated as a whole as if it were one individual species (whence the term 'quasi-', i.e., resembling a species). The results of this theory have now been confirmed by experiment. Charles Weissmann has cloned individual mutants of the virus Q_β and determined their sequences, thus obtaining information on the distribution of sequences in the mutants.[14] It was indeed found that the dominant sequence was present in only small numbers — so small, in fact, that none of the twenty clones investigated turned out to represent it perfectly (Vignette 9).

It also proved possible to confirm quantitatively the theoretically predicted relationship between error rate and quantity of information.[15] The dominant sequence is selected stably only if it possesses a selection value that is greater than the average efficiency of replication of the ensemble of mutants. If the incorrect copies are not to displace the dominant sequence altogether from the population, then the error rate must lie below a certain threshold value (Vignette 10). This threshold value of error rate is given approximately by the reciprocal of the number of information-bearing symbols in the sequence. In other words, the longer a sequence is, the more accurate its reproduction must be; otherwise errors accumulate in successive generations and the original information is lost. The genome of the Q_β virus contains 4200 symbols. According to Weissmann's experiments, the error rate of the replication *in vivo* is about 3×10^{-4}. There are thus, on average, one or two errors in each round of replication.[16] This means that the natural system operates just below the error threshold and produces the highest number of mutants that the system can tolerate. In this way, the system keeps itself stable, but is in a position to react flexibly to changes in its environment. It thus attains the highest possible rate of evolution. Investigations of this kind have now been carried out on a whole series of different RNA viruses.[15] They all show that populations of viruses are natural quasi-species distributions.

Crossing the error threshold causes instability. This is what happens when a selectively advantageous mutant appears. The sequence previously established no longer fulfils the threshold condition. It continues to make errors, and is no longer able to compensate for this loss by having an advantage over other competitors. It dies out, while the new variant grows up. Afterwards, the dice continue to be thrown and the game of evolution continues on the new, higher level.

Individual, exact results depend upon the error rate, the lengths of the sequences, the efficiency of reproduction in comparison with the average efficiency of the mutant distribution, and the size of the population. Evolution experiments by Sol Spiegelman and by Christof Biebricher have shown that natural processes indeed proceed in the manner we have described[15] (Vignette 11).

In making these measurements, the researchers subjected their system to constraints that were easily realized, and which thus allowed a quantitative test of the theory. If we choose to examine the evolution of species in Nature, we need to remember that the constraints and subsidiary conditions are often extremely complex. Nevertheless, at higher levels of evolution, selection (especially selection against incorrect copies of the same species) remains an inviolable law. Admittedly, the many additional conditions complicate the process to such an extent that quantitative predictions are generally impossible. 'Fittest' on the level of human beings is no longer a property that can be correlated with measurable characteristics. We must therefore be wary of extrapolation. Experience gathered at lower levels must not be projected carelessly on to higher ones. This kind of extrapolation has frequently obstructed the way to a correct understanding of the Darwinian principle, in spite of the fact that it is probably the most important principle of organization for the origin and development of life.

What has the selection principle told us about the origin of information? In answering this, we shall again restrict our consideration to the initial stages of evolution, namely, the origin of information in single molecules.

7. Evolution means the optimization of functional efficiency

... but life itself seemed without antecedent. If there was anything that might be said about it, it was this: it must be so highly developed, structurally, that nothing even distantly related to it was present in the inorganic world.

How good are our genes at their job? Their translation products are functional molecules, and these certainly fulfil with high efficiency their tasks as enzymes, as catalysts, and as regulatory units. But are they really optimal? And what do we mean by optimal? We can dissect the reaction mechanisms of enzyme-controlled processes down to the most elementary steps, which take place on the time-scale of one thousand millionth of a second. We can determine the highest rates that the laws of physics allow a given elementary step to attain. From this we can deduce how the individual steps must be adjusted in order to work in concert for the highest overall efficiency. In many cases, enzymic catalysis accelerates a reaction by a factor between one million and one thousand million. Wherever such a mechanism has been analysed quantitatively, the result has been the same: enzymes are optimal catalysts.[17]

The genes of the smallest proteins are made up of some three hundred basic units (nucleotides). Perhaps primitive genes were smaller still. Within the last few years, structural domains have been found in proteins, suggesting that the proteins may have originated by the coming together of smaller units. So primitive genes will probably have been some 100 to 300 nucleotides long. The number of permutations of a sequence of this length makes the number of alternative sequences so great that the chances of obtaining the optimal sequence by random arrangement are quite negligible.

Does the selection principle offer a solution to the problem of complexity? Can it be used to explain the origin of primitive genes and the optimization of their functional efficiency?

If we start from the popular interpretation of the Darwinian principle, we soon run into trouble. Today it is generally accepted that selection is an inevitable property of self-reproducing systems. Once an advantageous mutant has appeared in statistically significant numbers, these numbers will inevitably rise, and the new mutant will ultimately come to dominate in the population. Only at the beginning is there any uncertainty. As long as there is only one such sequence, or only a few, then it or they may be extinguished by some chance event before ever having the opportunity to reproduce. Deterministic selection always has a stochastic initial phase, and hence our allusion above to statistical significance. But the vital feature is our assumption that the mutant appears as the result of a *chance* event, a statistical fluctuation. In other words, each mutant appears with a probability that is independent of whether it is a superior, a neutral or an inferior variant. This in turn would mean that superior mutants appear only with extreme rarity. However, this interpretation reintroduces the problem of complexity, which was supposed to have been solved with the help of selection.[18] This may not seem clear at first reading, so let us take a concrete example. If a gene consists of 300 nucleotides, then it has some $4^{300} \approx 10^{180}$ possible sequences. Most of these are completely useless, a reasonable number will be partly functional, and only a very few will encode a given optimal function. In a random distribution of starting sequences, there will therefore at best be just a few, rather poorly adapted variants. Only through further mutation and selective stabilization of the superior variants, by prolonged iteration, can the best possible structure ultimately be reached. The greater the demands placed on it, the greater is the distance to the next, improved variant. This would be the result of a simple alternation between random mutation and inevitable selection. Yet it could only lead to the goal if the selection value rose uninterruptedly towards the optimum.

However, we know that selection values do not rise uninterruptedly. The distribution of selection values for the mutants in a given system should rather be seen as a landscape with plains, hills, and mountain ranges. The selection mechanism sees to it that the highest value peak represented in the initial distribution, even if just one or a few examples are present, becomes occupied by many copies, and that the mutants group themselves around this wild type, which is the best at the start and therefore the first to be selected. But now the system comes up against a dilemma. The value peak is not yet optimal. So it turns into a trap, from which the population distribution can only escape by a jump across to the next, higher peak. If the value peaks are distributed randomly, and the jumps are also random, then the localization of the mutant distribution by an initial selection among the sequences present would be more a hindrance than a help, since it would prevent the system from trying out all alternatives. Moreover, in the absence of any guidance, the system would have to try out all possible mutations in order to find an advantage.

This dilemma was perceived in Darwin's time, but its real extent becomes clear when the chances of mutation are estimated quantitatively. For our small gene with 300 nucleotides, the frequency with which a particular ten-error mutant occurs is no more than 10^{-30}. This jump of ten errors means a change in about three per cent of the gene. If we were to begin with a random sequence, we would probably need to replace about three-quarters of the symbols before reaching the optimal sequence. (The numbers in this example are based upon a presumed error rate of 3×10^{-3} per symbol and the fact that each symbol in the initial sequence has three ways of mutating.)

The assumption made in classical genetics that the production of mutants is completely random rests partly upon the fact that the mutation process takes place unseen. A mutant in a population cannot be observed until it has presented its credentials by expressing its superiority in phenotypic characteristics that are clearly distinct from those of the wild type. Population biologists, using abstract models, long ago reached the conclusion that many neutral mutants must exist, differing only slightly or not at all from the wild type. Motoo Kimura has shown that these neutral mutants spread with a probability independent of the size of the population and displace the earlier wild type. This drifting through mutant space is presumed to make an important contribution to the overall evolution process.[19] Most important, it could free the system from selection traps such as a value plateau lower than, but much broader than, the optimal value peak.

Later we shall see the importance not only of the neutral mutants, but also of those mutants present in the quasi-species distribution that, although somewhat less efficient than the wild type, are present in larger numbers than mutants only barely capable of survival. Such details seemed to be dispensable in the traditional theory. The classic phrase 'survival of the fittest' gives away the fact that only the wild type, the best-adapted individual type, was considered. Mutants were regarded as additional statistical fluctuations, necessary as a source of evolutionary progress, but for the most part neither observable nor identifiable. They could in no way be predicted or influenced. The exceedingly rare appearance of a superior mutant could thus be regarded as an independent and purely random event, and its rate of appearance accounted for by an *ad hoc* frequency. For the molecular biologist of today, however, mutants can be demonstrated by cloning and identified by sequence analysis. It is therefore now worth looking at the mutant distribution in detail. The theoretical approach must therefore include the exact registration of the fate of each individual sequence. When this book-keeping exercise was first performed, it led to a surprising result: one that demanded a complete reinterpretation of the Darwinian selection principle at the molecular level and presented the neutral mutants in a completely new light.

Theory and experiment show unanimously that, for viruses, the proportion of the population occupied by the true wild type is relatively small; the wild type appears macroscopically only because it is observed as an averaged sequence, termed 'consensus sequence', of the whole mutant ensemble. In the ensemble, the total number of mutants is usually much greater than that of wild-type individuals. Even if the wild type is the most common individual sequence, the number of all mutants taken together is far greater. The frequencies with which individual mutants occur depends primarily upon their degree of kinship with the wild type. The more distant this kinship, the smaller will be the chance of generating the mutant directly from the wild type. If all the mutants were to reproduce substantially more slowly than the wild type, then their individual frequencies would be given quantitatively by a Poisson error distribution.

However, the distribution is modified, sometimes drastically, if a mutant can replicate at a rate comparable to that of the wild type. This can sometimes mean that its population number rises to approach that of the wild type. The next consequence is that the mutant produces new mutants of itself. Finally, even mutants relatively distant from the wild type may appear in reasonable amounts, as long as the chain of their precursors replicates almost as efficiently as the wild type. An amplification of this kind can make a difference of several orders of magnitude (see Vignette 10).

The generation of a superior mutant, which was left to pure chance in the neo-Darwinian model, is now seen to be determined by the following causal chain.

- Mutants closely related to the wild type arise primarily by erroneous copying of the wild type itself. A distribution develops, in which close relatives of the wild type appear the most frequently.
- Selective advantage can generally be expected only for relatively large mutation jumps. These cannot occur frequently by statistical fluctuation.
- The initial, value-free distribution is modified by selection of preferred states within the quasi-species distribution. Functionally competent mutants, whose selection values come close to that of the wild type (though remaining below it), reach far higher population numbers than those that are functionally ineffective.
- An asymmetric spectrum of mutants builds up, in which mutants far removed from the wild type arise successively from intermediates. The population in such a chain of mutants is influenced decisively by the structure of the value landscape.
- The value landscape consists of connected plains, hills, and mountain ranges. In the mountain ranges, the mutant spectrum is widely scattered, and along ridges even distant relatives of the wild type appear with finite frequency.
- It is precisely in the mountainous regions that further selectively superior mutants can be expected. As soon as one of these turns up on the periphery of the mutant spectrum, the established ensemble collapses. A new ensemble builds up around the superior mutant, which thus takes over the role of the wild type.
- The occupation of the ridges of the value mountains by efficient mutants steers the process of evolution systematically in the direction in which a higher peak is expected. This circumvents the need to try out blindly a vast number of valueless sequences.

This causal chain results in a kind of 'mass action', by which the superior mutants are tested with much higher probability than inferior mutants, even if the latter are an equal distance away from the wild type. This is true in spite of the fact that the fundamental mutation step is purely random, or statistical, in nature. So there is no magical or prophetic force at work that steers the development of the mutant spectrum, but just the purely physical 'weight of numbers'. This directs the process straight into the mountain regions of the value landscape. The process is subject to the following rules:

Similarity in the sequence of the repeating units of the genes results in similarity in the sequence of the repeating units of the proteins that they encode. It is these proteins whose functional efficiency is judged by the selection to come. Similarity in the sequence of the repeating units of the protein chains results in similarity in their folded structures. Similarity in the folded structures of the proteins results in similarity in their functional properties and their efficiency. However, these relationships of similarity are by no means governed by a simple law of proportionality. Similarity of protein sequence need not in every case mean similar efficiency in function. There is merely a general, continuous trend, as in a mountain landscape in which peaks generally pass smoothly into valleys, but with occasional interruption by a steep face or a gorge.

Models that take account of this kind of behaviour have been investigated by theoretical calculation, by numerical simulation, and by experiment.[20] They lead to an abstract description that makes high demands upon our ability to visualize ideas. In a sequence with ν symbols, any of the ν positions can be mutated with much the same probability *a priori*. This is rather like starting from a given point in space and jumping in one of ν possible directions. The space that this picture describes we call *sequence space*. Sequence space is thus an abstract world with ν dimensions. Each of the ν coordinates in sequence space is a line of four discrete points that correspond to the four ways of occupying the position of a real nucleic acid. Mutation at a given position means jumping from one point to another along the associated coordinate. The number of points in this space is equal to the total number of all possible sequences.

The concept of a ν-dimensional space of points for binary sequences (those consisting of only two types of symbol, see Vignette 12) was introduced into information theory by Richard W. Hamming, and its application to the treatment of problems of evolution was first suggested by Ingo Rechenberg.[21] If we wished to find our way around such a ν-dimensional point-space, we would probably end up feeling like Alice in Wonderland. A binary sequence with 360 positions (corresponding to a relatively small gene) has $\nu = 360$ dimensions, and produces a sequence space that is simply immeasurable, and which would suffice to map the entire universe by the Ångström unit. In spite of this, only 360 steps are needed to cross the entire space. In this space, there are no great distances, but there are many twists and turns, and it is just as easy to get lost here as in the vast expanses of outer space. A characteristic property of sequence space is the extreme complexity with which routes passing through it are intertwined. Just wandering around aimlessly is of little help, since in the search for a particular position practically all positions would have to be looked at. However, guidance is provided by selection, which follows the value topography and thus reduces substantially the freedom of movement. For a connected range of value mountains, whose contours are not random but surround one or a few well-defined peaks, the peculiarities of the ν-dimensional space are of enormous advantage: this is due to the fact that, although the volume is large, the distances between points are small and their connecting routes are intertwined. Thanks to the way in which the system orients itself according to the value criterion, the target sequence appears almost to have been aimed at. Even so, the optimization is subject to certain constraints. First of all, the length of the sequences is restricted; secondly, an optimal error rate adapted to the length is required; thirdly, the population numbers must be high enough to guarantee a sufficiently widely distributed occupation of the value terrain.

Sequence space and quasi-species are two novel concepts, with the help of which the theory of molecular self-organization has been formulated in a quantitative way. It is a dynamical theory, in which population numbers of nucleic acids are time-dependent variables and their rates of change are adjusted by reaction velocity parameters.

The theory describes the origin of the information that is laid down in the sequences of the nucleic acids, and whose quantity is expressed by the lengths of these sequences. The information is needed for the optimization of self-replication under given environmental conditions. This information is coupled to its own origin in a feedback loop that is based upon the inherently autocatalytic nature of reproduction kinetics. This feedback loop is the ultimate cause of the self-organization that led to genetic information.

The new formulation of the concept of selection and its application to molecular systems differ sharply from the original Darwinian approach and from its later reformulation in population genetics. The target of selection is no longer the individual wild type that produces chance mutants in a completely random way. The object of selective evaluation is rather the entire ensemble of mutants, which we have denoted as a quasi-species. Here the dominant mutant (the so-called wild type) is still the most frequently represented individual sequence, yet its numbers may amount only to a minute fraction of those of the entire population. The sequence of the wild type can still be determined unambiguously in the laboratory, because it is normally identical to the average sequence (the *consensus sequence*) of the complete mutant spectrum. This fact has now been confirmed experimentally by cloning and analysis of individual sequences from various quasi-species, both artificially produced distributions of RNA molecules and the mutant spectra of natural viruses.[16]

There is an essential difference between the ideas expressed here and the neo-Darwinian idea of an alternation between mutation (= chance) and deterministic selection of the superior mutant (= necessity). In the neo-Darwinian view, the system is obliged to try out most of the possible sequences, and it can easily get stuck on a local optimization peak. The critical difference is connected with structure of the quasi-species. Since the quasi-species is evaluated as a whole, the individual population numbers for the mutants do not (as they do in the classical model) depend upon their distance from the selected wild type alone, but depend also upon their individual degrees of fitness and upon their distribution in the surrounding value landscape. Neutral or nearly neutral mutants are far better represented than others, because of their relatively efficient self-reproduction. If an efficient mutant lies in a region of efficient mutants, that is, in a mountainous region of the value landscape, then it will not only reproduce itself, but will also arise by the erroneous copying of its neighbours. This reinforcement leads to a dramatic shift in the population numbers. The usual decrease in number accompanying increasing distance from the dominant sequence becomes modulated; the *population topography* becomes a distorted picture of the *value topography*. The closer the fitness of a mutant comes to equalling the fitness of the dominant sequence, and the more sequences of equal value it is surrounded by, the better it is represented in the population.

The fact that the mutant distribution is asymmetric, in high-fitness regions, reaching far out into sequence space, has two consequences that we do not find in classical Darwinian theory, and which greatly bias the randomness of the generation of new mutants.

First, most mutants arise in the mountain regions of the value landscape, close to the high peaks of optimization. The system searches with high efficiency for mutants in the region where superior mutants are most to be expected. Narrow peaks are not even aimed at.

Secondly, aimless drifting is strongly restricted, because genuinely neutral mutants hardly appear. This does not mean that there is not a large number of mutants with fitness values almost equal to that of the wild type. But, since the selection decision depends just as much on an evaluation of the neighbourhood, two really equivalent mutants must not only have identical fitness values but also identical neighbourhood contours. This is highly improbable, so that macroscopic ensembles, which in general consist of 10^{10} sequences or more, will only very rarely possess cases of exact degeneracy. In the quasi-species picture, the concept of neutrality can be dispensed with. In classical theory, it has remained an open question how far two selection values can be allowed to differ and still be regarded as equivalent. In the quasi-species model, each sequence, regardless of how closely it resembles the wild type in its selection value, is evaluated exactly, together with all the sequences in its neighbourhood. Neutral theory in its present form refers to the limiting case of small populations and large genomes, where the appearance of any mutant is a unique, random event. The quasi-species picture, in contrast, requires finite and reproducible expectation values for population numbers. A general stochastic theory would unify both models.

Mathematically, selection can be formulated by an *extremum principle*. While this applies strictly only under certain conditions related to the mechanism of reproduction, it brings selection into line with other physical phenomena that also are characterized by extremum principles. The best-known example of these is thermodynamic equilibrium, characterized by a maximum of entropy or a minimum of free energy. On an abstract plane, selection means the localization of the mutant spectrum — the quasi-species — in a restricted region of sequence space. As sequences represent information, one sometimes speaks of a *condensation in information space*.

The error-threshold condition[22] is disobeyed every time a superior mutant appears. The existing quasi-species is destabilized; it 'evaporates', only to 'condense' again in a different part of sequence space. Evolution can thus be likened to a succession of phase transitions. Its movements are in no way aimless, or undirected, since superior mutants occur with greatest probability in the mountain regions of the value landscape. This kind of 'pre-programming' depends naturally upon the population size. It takes place at the microscopic level, but is still to a certain extent deterministic. This directedness of the evolution process is perhaps the clearest expression of the present-day paradigmatic change in the established Darwinian world-picture.

Anyone who is accustomed to base his world-view dogmatically upon Darwin, calling himself a Darwinist, will be reluctant to accept this new interpretation. He will counter with his own accustomed view, which, seen ideologically, certainly offers a possible alternative. However, our argument is a physical one, so that two things are important: assertions must be logically (in the end, mathematically) deducible, and the consequences of a theory must differ from those of other models, with differences that are experimentally testable. The mathematical buttresses are restricted to the 'if–then' behaviour of the system. They set out from fixed boundary conditions and show what must necessarily follow from these. To this extent, a mathematical model is not a direct representation of reality, but only an abstraction of reality. It is therefore important to ensure, by observation of the actual processes, that the real and the abstract boundary conditions do actually correspond to one another. Seen in this way, our interpretation says simply: *If* selection results from differing efficiencies of reproduction, *then* this occurs in the sense of the quasi-species model and not in the way envisaged by the classical wild-type model. *If* evolution occurs on the basis of natural selection, *then* it is value-oriented. The mathematical formulation now points up the limits of our generalizations. A presupposition for the 'if–then' behaviour is a sufficiently large population. Mathematical exploration tells us that this condition is satisfied by the population numbers ($> 10^{10}$) typical of molecules, viruses, and microorganisms. Another presupposition is a limit placed upon the length of the genome. It is this that allows a sufficient number of quasi-species states to be populated reproducibly. The realism of this presupposition has been tested by examining RNA viruses. With these and with several laboratory systems, experiments can be conducted under defined conditions, and so far these have confirmed quantitatively the predictions of the mathematically developed theory.

The chief criticism of Darwin's idea was directed against its supposed claim to explain all of evolution. However, the development of life, from molecular systems to human beings, has passed through many stages of organization, and, while some of these were Darwinian in nature, many were fundamentally different. Since the preservation of all living systems is based upon reproduction, selection plays a role at all levels. But selection is expressed in many different ways, sometimes as coexistence or even cooperation, and sometimes as competition and the often irreversible weeding-out of some forms of life.

Quantitative estimates and experiments on model systems suggest strongly that the kind of molecular optimization described here must have reached its natural limit at a sequence length of a hundred to a thousand repeating units. This implies that the optimization of individual genes must have taken place before their integration into a giant molecule, the genome. The reason for this lies in the genome's length. In a bacterial cell, a single gene represents less than a thousandth of all the information present, so that the error rate must have become adapted to the more than thousandfold longer genome. This much lower error rate, had it applied to isolated genes, would greatly have slowed down their evolutionary optimization.

Leibniz wrote in his *Theodice*: 'If, among all the possible worlds, none had been better than the rest [in our usage, optimal], then God would never have created one'. This statement certainly comes close to the mark at the molecular level of life.

8. What are the natural prerequisites for the origin of life?

Sooner or later, division must engender some kind of units which, though conjoined, were not yet organized, and which mediated between life and non-life; molecular groups that embodied the transition between vitalized organization and mere chemistry.

We have seen that the origin of information, and of life, by selection requires *self-reproduction* and *mutation* far from equilibrium. These open up the way to a level of organization that rises in step with the ever-increasing functional efficiency of macromolecular structures.

But how did the first self-reproducing molecules originate? This is naturally a question that requires a historical answer. But here it is not quite so easy to separate chemical history from chemical principles. Our question is, in essence, 'Was the chemistry of our early planet rich enough to generate the basic chemical building-blocks needed for life?' These building-blocks will of course include the naturally occurring amino acids (the monomeric sub-units of the proteins); they will include the bases A, T/U, G, and C along with the phosphate esters of the sugars ribose and deoxyribose (the ingredients of nucleic acids); and they will include carbohydrates and fats. In short, the whole palette of 'organic' chemistry is called for. To answer our question with a downright 'yes' would be to cast doubt upon its fundamental nature: if we were to try to justify our 'yes', we would have more answers than we need. We know in many cases how things *could* have taken place, but not how they *did* take place. It is therefore no surprise that schools of thought have crystallized that sometimes resemble ideologies more closely than they do scientific scrutiny.

The first question to tackle is that of which came first, the proteins or the nucleic acids — a modern version of the Scholastics' chicken-and-egg riddle. There is no doubt that proteins, which are more easily formed, were the first on the scene. But how could these proteins be instructed to become optimal catalysts? In 1954, Stanley Miller, then a student of Harold Urey, showed by his classical experiment on prebiotic synthesis that amino acids are formed under a wide range of conditions, from a mixture of simple basic substances such as water, nitrogen, ammonia, hydrogen cyanide, and methane.[22] An important condition is that energy is supplied to the system, in a form such as electric discharge, shock waves, or high-energy chemicals. The convincing thing about these experiments is not so much that amino acids are formed at all; it is that these relative abundances correspond to those found in Nature in meteorite rocks, and the amino-acid compositions of proteins in living organisms show the same relative abundances. Sidney W. Fox[23] has gone one step further, and shown that separate amino acids can condense to give protein-like substances under conditions that were easily realizable on the primitive Earth. His proteinoids have another remarkable feature: they show a wide variety of catalytic ability, albeit extremely weak.

It was therefore possible for many different reaction products, of high and of low molecular weight, to form and to accumulate in large quantities in the oceans. It was equally possible that these substances included a few that possessed reasonable catalytic efficiency, especially at contact interfaces. Today we know that the immune system can make specific, closely fitting antibodies to fit any molecular shape. (An antibody is a protein with a high binding affinity for particular chemical patterns.) This fact implies that, even in the prebiotic phase, at least some proteinoids could be formed to catalyse virtually any given chemical reaction. However, these catalysts were not optimal, nor was it possible to optimize them, since optimization calls for a stepwise evolutionary adaptation based upon self-reproduction. Nevertheless, it seems probable that some of these proteinoids possessed a certain stereospecificity, that is, they were probably able to distinguish between left-handed and right-handed molecules, or *enantiomers*. Our problem is to find out, among the many possible routes evolution could have taken, which route it actually took.

Similarly, there are many chemical pathways that could have led to the first nucleic acids, especially when we remember the existence of catalytically active proteinoids. Leslie Orgel and his co-workers[24] have performed an impressive series of experiments in which they showed that, even under simple conditions, nucleic acids are capable not only of polymerization from simple monomers, to give chain-like molecules containing over 100 monomer units, but also of self-replication, without the help of enzymes, by virtue of their own template-like structure. This self-replication is catalysed potently by metal ions such as magnesium and zinc. Interestingly, it is precisely these ions that today assist the action of many of the enzymes that make DNA and RNA by copying from a template.

The search for a prebiotic catalyst for RNA synthesis, a primitive polymerase, is perhaps one of the most important aspects of the question asked in the title of this chapter. Present-day polymerases are complex enzymes, most of them assembled from several different sub-units, and they represent final, optimized products of evolution. They show not only that a stereospecifically correct synthesis of nucleic acids from energy-rich monomers is possible but also that it can be achieved by means of relatively few commonly occurring active groups, as long as these can be arranged correctly at the active centre of a protein molecule. The time needed to 'copy' one monomer is between a hundredth and a thousandth of a second. It follows from this that there must be a vast number of simpler catalysts that could perform the same task, though of course with lower efficiency. It is, for example, highly interesting that single-stranded ribonucleic acids can fold up spontaneously in such a way as to engender efficient catalytic functions that may include polymerization; Nature still employs these catalysts today. They are frequently referred to as 'ribozymes', a term cannibalized from 'ribonucleic acid' and 'enzymes'. The reaction known to be catalysed by ribozymes is a tailoring of other nucleic acid molecules, termed *splicing*, a reaction that follows much the same chemical rules as does polymerization. Polymerases of this kind are also known to be adapted to the tasks of breaking down, or hydrolysis of nucleic-acid chains. Does this mean that the nucleic acids were able to evolve without the support of proteins or proteinoids? This historical question cannot yet be settled, but we are at least certain that there exist many possible ways of influencing catalytically the polymerization reaction. Even among stereospecifically active catalysts, there may be some, perhaps purely inorganic, precursors that select energy-rich monomers from the jumble of molecules around them and join these up into a polymeric chain, as the enzymes do, but less well. The frequencies of the different classes of polymers that arise cannot be deduced from equilibrium measurements, as the reactions occur far from equilibrium and depend critically upon the exact conditions of the environment. As in 'real life', the polymers are merely metastable, short-lived intermediate states, whose ultimate fate is hydrolysis by their aqueous environment and reduction to energy-deficient monomers. In just the same way, all living entities, from the earliest precursors to present-day organisms, are dynamic systems far away from chemical equilibrium.

There are known chemical pathways, leading to the synthesis of the various components of nucleic acids, that could well have been realized under prebiotic conditions. The simplest of them lead to the synthesis of adenine, the base A, which can, for example, be made by the mere condensation of hydrogen cyanide. The base G (guanine) can be regarded as a rearranged oxidation product of A, and thus as having arisen by direct route. Synthetic paths for the two complementary pyrimidine bases, C (cytosine) and U (uracil) with its congener T (thymine) are harder to work out. Up to now, U has only been made under natural conditions by the oxidation of C, for which a possible prebiotic synthetic pathway has been found.[25] In non-aqueous reaction media there exist further, elegant synthetic procedures for all four bases, as Albert Eschenmoser has shown.[25]

It is virtually certain that the four monomers existed, during the early phase of synthesis, in widely differing concentrations. The easy synthesis of A suggests that there was a great excess of this base. However, the accumulation of large quantities of A-rich polymers would have been restricted by the initial shortage of the complementary base U; this is because A in a template strand is copied as the complementary U in the daughter strand, which in turn is copied as an A, and so forth. This process is analogous to photography, in which a picture is copied by way of its negative (see Vignette 4). Bases G and C may well have existed in a more balanced ratio and thus enjoyed better starting conditions for template-instructed replication. The GC pair, held together by three hydrogen bonds, is about ten times stabler than the AU pair with only two. The stability of complementary base pairing plays an important role, as the newly formed strand must remain attached to its template until the entire message has been copied.[26]

Calculation of the necessary binding energy, along with experiments on binding and synthesis, show that sequences rich in G and C are best at self-replication by template instruction without the help of enzymes. There is also historical evidence for this. The precursor sequence for transfer RNA can be reconstructed, and it is found that this had a very high content of G and C.[5] Transfer RNA (tRNA) is one of the 'oldest' molecules found in present-day cells. Its composition has changed little in the course of evolution. There are molecules of tRNA for which the sequences possessed by humans and by the frog *Xenopus laevi* are identical.

At the same time, examination of the genetic code (Vignette 8) indicates that its first codons were rich in G and C. The sequences GGC and GCC code respectively for the amino acids glycine and alanine, and because of their chemical simplicity these were formed in greatest abundance in Miller's prebiotic experiments. The same applies for the next two assignments, GAC and GUC, which follow from the reading frame defined by the first and third letters in the three-letter code-word. The assertion that the first code-words were assigned to the most common amino acids is nothing if not plausible, and it underlines the fact that the logic of the coding scheme results from physical and chemical laws and their outworkings in Nature.

Another remarkable feature of the code is the structural and logical basis that it provides for the transmission of information. Because of the different abundances of the bases, complementary copying is essential for the storage of information. If the system were to reproduce itself directly, then it would come to be dominated by uniform polymers of the most common monomer, and information cannot be stored with the help of an alphabet that contains one letter only. Complementary reproduction assures that the second symbol is introduced, and also that ultimately both symbols are represented in the polymers with (on average) equal frequency. An excess of one of them in one strand results in an excess of the other in the other strand. This leads to an inevitable tendency towards a uniform mixture. Over and above that, kinetic factors are responsible for a selection pressure that keeps the two strands as nearly identical as possible.

One might well ask why Nature has used four symbols, when she might just as well have made do with two. As we know from our experience with teleprinters and computers, two symbols are quite sufficient for the storage and transmission of information. Machine code can get by with so few symbols because machines use man-made devices for transmission and reading. In the early period of evolution there were no such machines. The selective processing of signals required the various nucleic-acid sequences to have different structural stability. With only one complementary base pair, all sequences might well be stable, but their stability would also be largely homogeneous. They would hardly offer distinguishing characteristics for an information-processing molecular machinery that had to come about by self-organization. With two complementary pairs, on the other hand, the many combinations possible would open up a broad palette of differing stabilities, and thus provide a vast and expressive repertoire of distinguishing characteristics. Analogous assertions can be made about the construction of the code-words and their assignment in the genetic code.

There are naturally many questions that still remain open, and not only those concerning historical boundary conditions. For example, on what level was the handedness, or chirality, of biological molecules decided? We know that the proteins made by the information-instructed cellular machine use exclusively the 'left-handed' amino acids, and thus produce left-handed structures. The nucleic acids employ 'right-handed' monomers, which depending on circumstances can adopt right- or left-handed spiral forms.

It is understandable enough that a selective mechanism, starting from a random mixture, will choose one or the other extreme in preference to a compromise, simply because a machinery committed to one kind of handedness is easier to construct than one which is required to 'change hands' again and again. Since life is the result of an evolutionary process that in certain phases had to make a decisive, once-and-for-all choice, it comes as no surprise that the chirality of the macromolecules in all organisms is the same. But is there in fact any sense in asking why in one case 'left' and in another case 'right' was selected? Even if there is no *a priori* physical explanation for the decision, even if it was just a brief fluctuation that gave the one or the other equivalent possibility a momentary advantage, the self-reinforcing character of selection would turn the random decision into a major and permanent breach of symmetry. The cause would be a purely 'historical' one.

How could the first macromolecular sequences come about at all, in a chemical environment that was racemic (i.e., that contained right- and left-handed forms in equal amounts)? Enzyme-free replication experiments with nucleic acids have been carried out to see whether right-handed templates can give rise to left-handed products. All results obtained so far have shown that like-handed copies like-handed, and the presence of monomers of the opposite hand inhibits the reaction. A possible answer is that there were perhaps primitive stereospecific catalysts that always chose one or the other form, while later selection did the rest. Not only was the world enriched with proteinoids and other complex chemicals, but the surfaces of minerals also may have played a role in selection. Another explanation, perhaps less plausible, is a break in symmetry on the level of small molecules that — at least in certain regions of the Earth — provided a chirally pure environment. A third possibility, with a sound enough chemical basis, was suggested independently by Albert Eschenmoser and Leslie Orgel: the chiral nucleic acids might have had achiral precursors. The chiral component of nucleic acids is the sugar (ribose or deoxyribose). It is conceivable that these arose, after a period of evolution, from an achiral compound. With complementary bases, an achiral compound would have been able to reproduce, and therefore to be subject to the rules of selection and evolution. In this view, today's forms are an optimized end-product of chemical evolution.

Here again, there are too many answers rather than too few. We are not confronted with a paradox for which we cannot find an explanation. The problem is an *embarras de richesse* of physical and chemical explanations. Although research groups the world over are working on these problems, only a few of the possible mechanisms have been examined by experiment.[27]

The chemistry of early evolution is still a determining factor for all life on our planet. It is, in fact, surprising how conservative evolution has been with regard to chemical mechanisms. This is seen most clearly in cases where changes have been imposed by internal or external (i.e. environmental) conditions. A good example of this is the limitation placed upon the information content of single-stranded sequences by the limited accuracy of copying. The change-over to double-stranded sequences made it possible to proof-read the copies after they had been made, which led to an increased capacity for information. Chemically, this was expressed in the change from ribonucleic acid (RNA) to deoxyribonucleic acid (DNA). Even today, both the metabolism of DNA and its reproduction still bear witness to this change. Sol Spiegelman[28] once remarked: 'The nucleic acids invented human beings in order to be able to reproduce themselves even on the Moon.'

A classic example of a modification of chemical mechanisms imposed from outside is the change from aerobic to anaerobic metabolism, which was caused by a change in the atmosphere's composition some two thousand million years ago. While there is no unanimity about the exact composition of the early atmosphere, there is certainly agreement that it did not contain free oxygen: so it was not oxidizing. Such oxygen as there was existed as part of the compounds water (H_2O) and carbon dioxide (CO_2). It is harder to decide to what extent the atmosphere was *reducing*, that is, how much free hydrogen it contained. The earliest organisms had to extract the energy for their metabolism from the fermentation of energy-rich compounds of inorganic origin. This meant that the improvement of the energy supply was one of the primary demands on evolution. This selection pressure was responsible for the appearance of anaerobic photosynthesizing bacteria, which exploited efficiently the energy of the sun. They were followed by the cyanobacteria, which also carried out photosynthesis, and were in turn the forerunners of the chloroplasts, which today are the power-houses of green plant cells.[29] Over a long period, the cyanobacteria released large quantities of oxygen. This oxidized the iron that was dissolved in sea water, and the oxidized iron precipitated as huge band structures in rock that today are the evidence of the oxidation process. Only gradually did oxygen gas begin to spread throughout the atmosphere, which previously had been kept oxygen-free.

There has therefore been a phase during which the environment was just as much affected by life as life was by the environment. The appearance of oxygen in the atmosphere led to the formation of the ozone layer that today filters out ultraviolet light and is indispensable for the existence of present-day life forms. The free oxygen was ultimately not only tolerated but also put to use in an improved metabolism. Chemically, respiration can be regarded as photosynthesis put into reverse, consuming oxygen and producing carbon dioxide and chemical energy in the form of adenosine triphosphate. It is true that the evolution of aerobic metabolism called for many new reaction paths, such as the citric acid cycle; yet, even so, they were no more than modifications of paths already in use. They made use of pre-existing possibilities by exploiting substances that, in the old fermentation routes, were mere waste-products. They used the same starting materials (carbohydrates) and stored the metabolic energy gained from them in energy-rich phosphates. These mechanisms prevailed because they were incomparably more efficient and made better use of available resources. The energy yield of respiration is eighteen times greater than that of fermentation. The evolutionary kinship of these processes is seen in the fact that many aerobic cells are still able to switch back to fermentation-based metabolism if they become starved of oxygen.

In this chapter, we have been concerned with the environmental conditions necessary for the origin of life. We can state such conditions and investigate them in the laboratory. However, we cannot know whether the processes that we postulate are the real historical ones, or whether there are yet further mechanisms, of which we know nothing, that lead to the same results. Thus, many of the historical details of the origin of life may still be unrevealed.

The requirements for metabolic turnover in the living cell are very complex. That is, the processes are very susceptible to disturbance. Were there ever disturbance-free conditions in the prebiotic environment under which evolution could start up? Again, we are asking a historical question. In order at least in principle to answer with a 'yes', we would have to be able to point to catalytic mechanisms that can work under realistic, natural conditions. Many laboratories are working hard on these problems.

What about disproof? An argument of this kind — like, for example, the 'proof of the impossibility of perpetual motion' — cannot be produced in the case of the evolutionary origin of life. It has been tried, but with false premises, namely, the assumption of a perpetual chemical equilibrium. Life is a notorious non-equilibrium phenomenon! The prebiotic processes that led to the origin of genetic information had to occur far away from chemical equilibrium of the information-storing nucleic acid molecules, because otherwise hydrolysis would have wiped out the stored information. The polymerization of nucleic acids away from equilibrium can be reproduced in the laboratory in a relatively simple experiment.

In order to deny the possibility of a natural origin of life, one would have to be acquainted with all the historically possible conditions and then to show that there is no catalytic mechanism that under *any* of these sets of conditions could have carried out the desired task. Such a proof is hardly conceivable, on account of the enormous number of possible mechanisms and conditions that would have to be excluded. But the possibility of such a proof can itself be disproved. The biosynthesis of the living cell is already a mechanism of the kind in question. It is admittedly complex, but is completely interpretable within the scope of our present-day physical and chemical knowledge. This does not mean that everything in this realm has been investigated exhaustively; instead, it means that what has been investigated has turned out not to be mysterious, but to be explicable by physics and chemistry. What we know makes it probable that there are also simpler, less efficient mechanisms that were realizable under prebiotic conditions. The reconstruction of sequences of nucleic acids from the beginnings of the origin of life, described in detail in Chapter 2, implies strongly that the optimization of the biosynthetic repertoire took place in an early phase of evolution and within a relatively short period. The historical evidence that every cell bears within itself points to a continuous evolution of molecular mechanisms. The chemistry that we find in living organisms is in principle identical with the chemistry that we practise in our laboratories.[30]

9. The ladder of organizational levels[31]

Between the protean amoeba and the vertebrate the difference was slight, unessential, as compared with the difference between the simplest manifestation of life and that Nature which did not even deserve to be called dead, because it was inorganic.

Our discussion up to now has led us to three conclusions.

- Fossils and other relics of the development of life on Earth suggest that it originated gradually, beginning about four thousand million years ago.
- The molecular construction and the reaction pathways of living cells are in harmony with our general knowledge, based on physical theory and chemical experiment, of how matter behaves.
- Our insight into the physics of material behaviour enables us to deduce mechanistic principles that make comprehensible the origin of a form of organization, based upon optimal functional efficiency, that we can regard as an early form of life.

However, there is no 'universal formula' that could be used to deduce rigorously the origin of life and to explain its miraculous variety, from the simplest virus to the human mind. On every level of evolution, we can identify mechanisms that allow the system to develop further. They have much in common and differ greatly in detail. Our task is not just to build up certain chemical structures; a self-organizing system also needs, so to speak, to be 'motivated' to develop itself. Ultimately, it has no choice but to adapt itself to the given environmental conditions, for the creation of which it itself is partly responsible.

We have seen that the most important principle in this kind of self-organization is that of natural selection, as enunciated by Charles Darwin and Alfred Russel Wallace. It can be recognized, in various modifications, at many different levels of organization. For simple replicators, it has the character of a natural law (a simple extremum principle). These systems include not only self-replicating macromolecules of DNA and RNA, but also more complex DNA- or RNA-containing units such as viruses or even autonomous, vegetatively reproducing microorganisms. We have seen that the prerequisites for this kind of evolutionary behaviour are the capacity for self-reproduction, the appearance of mutants that also can reproduce, and the fuelling of reproduction by the consumption of chemical or other energy. Self-reproduction (including complementary reproduction) is a process whose rate goes up fairly precisely in step with the size of the population. If unrestricted growth can take place, then there is an exponential increase in the number of individuals. If the rate of reproduction were not proportional to the population number, but instead had some constant value, then the result would not be selection but coexistence: the different species in the population would be represented by numbers proportional to their respective growth rates. The best-adapted mutant would thus possess the greatest number of copies, but it would not be able to displace its competitors. When this situation arises in biology, of molecules or of organisms, it is referred to as *niche formation*.

If, in contrast, the rate of reproduction depends upon the population number in a manner that is stronger than linear, then the growth rate can become hyperbolic. If the total population number is restricted, this means that the ability of an individual to replicate itself depends not only upon its inherent replicative ability but also upon the number of other such individuals present. In this case, a new mutant cannot displace an established population, even if its properties make it superior, since the new mutant only appears in one or a few copies. The established form has therefore been selected 'once for all time'. The variety of species that inhabit our planet is incompatible with this kind of selection on the scale of the whole planet. However, the situation may occasionally occur on a smaller scale. A hyperbolic growth function differs from an exponential one in that it can reach an infinite value in a finite time. Since a world of limited resources cannot cope with limitless population growth, this situation must lead to a population catastrophe unless a way out is found first.

We have seen the principle that every species whose individuals can copy themselves, be the species DNA, RNA, or a microorganism, will obey the law of natural selection. But how do we get from a molecular replicator, the RNA or DNA molecule with its restricted information content, to a cellular replicator, such as a bacterium or a blue alga? What becomes of the selection principle when cells join up as organisms that do not reproduce vegetatively but sexually, as in the cases that Darwin had in mind? What happens when the replicator (the genotype) is no longer identical with the phenotype that is to be evaluated by selection?

We can see this dichotomy between genotype and phenotype even at the level of a simple object such as a virus. It typifies in many ways the pre-cellular phase of evolution, which has to pass through the phase of molecular replicators before reaching cellular replicators, even though the viruses we know are presumably products of a post-cellular evolution.

Let us take a fresh look at the process of viral infection, using as a simplest example the infection of a bacterial cell by an RNA phage (Vignette 13). The genome of the phage consists of a single-stranded RNA molecule that encodes usually no more than four different functions. What are these functions? First of all, the virus needs a material in which to pack and protect its own genetic information. Secondly, it needs a means of introducing its information into the host cell. Thirdly, it requires a mechanism for the specific replication of its information in the presence of a vast excess of host cell RNA. Finally, it must arrange for the proliferation of its information, a process that usually leads to the destruction of the host cell.

This programme can be realized by relatively few genes, each of which codes for a protein with one of the functions in question. The genetic information is translated with the help of the machinery of the host cell. The virus even gets the cell to carry out its replication; its only contribution is one protein factor, specially adapted for the viral RNA. This enzyme does not become active until a 'password' on the viral RNA is shown. When it sees this, it reproduces the viral RNA with great efficiency, while ignoring the very much greater number of RNA molecules of the host cell. Consequently, the cell is soon flooded with viral RNA. This is packed into the virus' coat protein, which is also synthesized in large quantities, and finally the cell bursts and releases a multitude of progeny virus particles.[32]

All this is a programme that runs automatically and is rehearsed down to the smallest detail. But how did it come about? The virus could not evolve until its host existed. Perhaps it was once a part of the cell that went its own way and modified its genotype so that there emerged a competitive phenotype. This seems particularly likely to have been the case for the virus-specific factor, the reproduction enzyme. The evolution of this factor called for a specific feedback between the product of translation and the viral gene. The translation product had to let its gene know, by influencing the rate of reproduction, whether a mutation in the gene was advantageous or disadvantageous for the phenotype. In this way, a replicase could arise that replicated exclusively viral RNA with high efficiency.

A comparable situation existed in the early phase of evolution during which the replicators began to translate their information into functional proteins. It is easy to imagine that short-chained, RNA-like replicators existed during this phase and functioned both as genes that coded for the amino acids in the proteins and as the adaptors that introduced them. In their function as adaptors, molecules of tRNA ultimately became incorporated into the genomes of cells. Many of their characteristics that can be identified today echo this earlier double role.

The analogy between the origin of instructed protein synthesis and the origin of the RNA viruses runs as follows. The host cell, with all its constituents, is replaced by an environment rich in chemicals and, especially, in more or less specific mechanisms for the translation of RNA sequences into protein molecules. This would bring no benefits in an evolutionary sense unless some of the translation products had a beneficial effect upon the reproducibility of their *own* particular genotype. Such a feedback would result in the selection of this genotype. Many possibilities for feedback, both positive and negative, are known. They include straightforward chemical effects such as the enhancement of a metal ion's catalytic power by the binding of a protein; more complex promotor or repressor effects; and, most specifically, enzymic catalysis that recognizes and selectively copies individual RNA sequences if they possess a specific structural element by which they can be recognized. The rate of replication in such a case of feedback will depend upon concentration in a manner stronger than linear, since the replicator is not just a template for its own reproduction; it includes both the template and the catalyst for the replication reaction. Both processes run at a rate proportional to the number of templates, so the rate of the process as a whole is doubly dependent upon this concentration: once because of the template itself, and once because of the catalytic function engendered by its translation. This strong dependence leads to hyperbolic growth, and, if population numbers are restricted, to once-and-for-all selection. A system containing a template-instructed replication cycle over which a catalytic feedback loop is superposed we call a hypercycle (Vignette 14). Very recently, such a hypercycle, with all its consequences for growth, selection and proliferation, was shown experimentally to be operating in the mechanism of infection of host cells by viruses.[32] In the example just discussed, the feedback loop includes only the two complementary strands of RNA, but such loops could easily integrate further sequences into a single cycle. This also occurs spontaneously, because every sequence produces mutants, and the mutants produce translation products that differ in small ways; these in turn can join in and build up further coupling patterns, so that the whole becomes a complex reaction network. In this kind of network, cycles arise and selection causes their components to pervade the population. The hypercycle arises automatically, by selection, whenever the feedback described above is present.

The RNA viruses are natural hypercycles. There also exist more complex viruses in which several individual RNA molecules, coding for products with complementary functions, are present simultaneously. A typical member of this group is the influenza virus, which houses eight RNA sequences, of which three encode various replication enzymes. Perhaps Nature has further, even more complex, hypercycles in store for us to discover.[32]

In the early phase of evolution that led to the autonomous cell, there were naturally far more problems to be solved than in the evolution of viruses. The construction of the translation apparatus alone calls for a whole set of genes to co-operate one with another. Because of the error-threshold relationship, which puts a strict limit on the length of stably reproducing nucleic acids, they could not be strung together into a long, continuous molecular chain. An integrated genome, which is what such a chain would be, would require an optimized machinery of reproduction, operating with a very low error rate, and this in turn would require an optimized apparatus of translation.

So, in the initial phase, somehow, the different genes were forced to co-operate with each other. For this to happen, the following criteria had to be fulfilled.

- First of all, each gene needed to protect its information from the accumulation of errors. To do this, it had above all to remain superior to its mutants.

- Secondly, the genes that represent the different, complementary functions were not to compete with each other. They needed to co-operate, and to allow their concentrations to attain stable values.
- Thirdly, the entire ensemble had to be capable of expansion and optimization. In this way, it was in a position to compete as a single entity against other ensembles of genes.

The only known form of organization that simultaneously fulfils all these conditions is the compartmented hypercycle. This can build up a regulatory system and, by virtue of its inherent feedback, integrate all its genes into a unified whole. The individual genes are preserved as individuals. In this connection, a recent discovery is of particular interest: the catalytic activity of RNA molecules. 'Ribozymes'[33] would be ideal components of simple hypercycles, especially if they could catalyse the formation, cleavage, or rearrangement of nucleic acids. In this way, information needed for the development of a translation system, such as a set of coding assignments, could accumulate slowly and be integrated piece by piece. The genetic code and the apparatus of translation could thus have appeared gradually, hand in hand.

However, this has not yet overcome the genotype–phenotype dichotomy. An advantageous mutation in the translation product of a gene (i.e., in the phenotype) cannot directly affect the fate of its own — mutated — genotype in the hypercycle. It has no selective advantage over its precursor gene.

Only those mutations that improve the RNA as the target of the feedback are recognized and optimised by the evolving hypercycle. In contrast, mutations that improve the coupling factors or their functions have no effect in selection. The hypercycle is indeed able to establish and reinforce a feedback loop by virtue of its rapid self-organization, but it remains functionally inefficient unless it finds a way to exploit the improvements in the translation products as well. This kind of feedback loop comes about when the components of the hypercycle are enclosed in a compartment. Here, the close proximity between genotype and phenotype brings about direct contact and acceleration of the reaction promoted by feedback.

It is reasonable to ask whether, in view of this, the hypercycle is really needed at all. Would not a compartment suffice to make the different genes work together? The answer is *no*. The genes would compete relentlessly within the compartment: even if they were all replicated by the same enzyme with same speed, then there would still be neutral selection among them, as described above. In addition, the genes are represented by a host of mutants with translation products of varying degrees of efficiency. There would therefore be compartments of differing efficiency and no selective stabilization. The compartments would degenerate, owing to the perpetual breach of the error-threshold relationship. Further evolution requires not only inclusion in a compartment but also the regulated co-operation of the partners[34] (see Vignette 14). And the recent discovery of a natural hypercycle suggests that, in addition to the enforcement of replication by positive feedback through replicases, various negative feedback loops with regulating power might also be involved. The promotor–operator–repressor regulation of DNA replication in bacterial cells provides a good example of this.

We have now come well along the way from the molecular to the cellular replicator. Indeed, two principal properties of the biological cell have been realized: its demarcation from the outside world and the regulation of its reaction processes. We are still in the dark about the details of the origin and the optimization of the cell's apparatus of biosynthesis, but many possibilities are under debate among researchers. This is a good place to sketch the 'if–then' logic of the construction.

An integration of all genes into a giant genome occurs only when the cell's enzyme kit has developed so far as to allow the error rate in copying to be reduced. For microorganisms, an error rate below 10^{-6} to 10^{-7} must be achieved. Without proof-reading, such small values can never be attained. This led to the need for double-stranded replicators. The parent strand served as a standard of comparison for correcting errors in the daughter strand. The phase of transition from the molecular to the cellular replicator was complete when not only all the genes had been united to give a genome but also the doubling of the molecular replicator had become synchronized with the division of the cell. This was the point at which the cell in its entirety, by virtue of the chemically based reproductive ability of the nucleic acids, became a higher-level replicating unit.

A new era of evolution could then begin, one in which selection in the Darwinian sense reintroduces the possibility of variability. The hypercyclic organization of the complex reaction network of nucleic acids and proteins, with its conservative 'once-for-all-time' selection, can now be seen as an intermezzo in the symphony of evolution, today fixed in the internal structure of the cell's regulatory mechanism. It is probable that this phase of evolution of the cell's replicatory mechanism was responsible for the universality of the chemical construction of the cell, for the universal genetic code and for the break in symmetry that imposed a uniform chirality upon every class of biological macromolecule. The integration of genetic information in the proto-cells (*procytes*) also left its mark in the operon structure, in which genomes are built up. The dynamic organization of the hypercyclic stage of evolution is reflected in the 'linguistic' order of the genome and the regulation of the cell's information retrieval.

Yet the centralization of the legislative in a single, huge DNA molecule also carries disadvantages. The integration could not take place until the error rate in reproduction was small enough to protect the giant molecule from an error catastrophe. But a small error rate makes the single gene almost immutable. The genome of the procyte contained the strictest economies in respect of information content. Their sequences were free of superfluous stretches, like the sequences of most present-day viruses, which economize by having genes that overlap. The evolution of a single gene that contains only a thousand nucleotides would, if the error rate was, say, one in a million, proceed very sluggishly. It is true that new mechanisms arose to allow some degree of variability, but the cell type of the procyte remained much the same. Changes in the genome are here handed down exclusively to the immediate descendants, the cell line.

For this reason, the future belonged not to the procytes but to another cell type, the eucytes. These are defined as cells that possess a cell nucleus surrounded by a membrane. A common origin for pro- and eucytes is indicated by structural relationships between them, and in particular by clear homologies in some of their sequences. Perhaps a precursor of the archaebacteria, by combination with a eubacterium, was the stage of life at which the pro- and the eucytes went their separate ways, more than two thousand million years ago. The cytoplasmic components of the eucytes show closer kinship with the archaebacteria than with the eubacteria. Their mitochondria, on the other hand, can be regarded as descendants of the eubacteria. The new cell type thus seems to have arisen from a union of two dissimilar precursors.

Today there are still many unicellular organisms among the eucaryotes. A typical one of these is yeast. While the genome of the procytes reflects an order that indicates a seamless integration of replicators, the genomes of eucytes reveal no such coherent gene structure, to say nothing of higher-order structures such as operons. The genes are divided up into segments, seemingly at random. The information-carrying portions of the genes are termed *exons* (expressed regions). These are interrupted by non-coding sequences, called *introns* (intervening regions). The function of introns, if they possess one, is still unknown. Neither is it known definitely whether they represent a relic of times past that perhaps could yield information about the origin of the eucytes. It has been found that exons are frequently correlated with structural domains in protein molecules. These domains, rarely more than a hundred amino acids in length, could have been primitive forerunners of the protein in question. Perhaps the introns became interposed when the genes were integrated to give a genome. They would then have vanished again in the procytes because of the economization of information content that took place; in the eucytes, they may have been stabilized by the acquisition of a function. According to another view, the introns were a characteristic acquired later in evolution that was found advantageous for the recombinative mechanism of heredity. The luxury of the accompanying expansion of the genome may have become possible when the error rate was as low as in present-day organisms, that is, less than one in a thousand million.

Recombination (Vignette 15) is the basis for the sexual mechanism of inheritance characteristic of the eucytes. Its essence is an exchange of sequence segments of DNA between the male and the female genetic material, so that in every generation new combinations appear and provide continual innovation. Apart from this increased variability, the most important consequence of this new kind of propagation is the shift of the front line of attack of evolution from the individual cell line to the population as a whole, whose gene pool now directly absorbs every change. With this move, the genes are *de facto* liberated from the central control of their genome. A selective advantage can now spread out horizontally over an entire population, while at the same time the angle of attack of a mutation harboured by one individual (or many communicating individuals) in the population is greatly widened. Nevertheless, this progress also had its price. Programmed death could no longer be avoided, or, put more directly, the ageing and death of the individual became so advantageous for the development of the species that they became an integral part of the evolutionary process. For organisms with vegetative cell division, the individual does not age. After division, the mother and daughter cells are indistinguishable. At this level, the only cause of death is mishap. So cells with vegetative reproduction are in principle immortal. Organisms with sexual reproduction, on the other hand, have progeny that are clearly distinguishable from their parents. It is advantageous that the individual, having made its contribution to evolution, dies. Its death means new life for the species.

The selection process with sexual reproduction is subject to 'social' problems similar to those encountered in the hypercyclic phase of evolution at the level of molecules. But the consequences here are different. Again, the simple extremum form of the selection principle is violated, and, again, account has to be taken of the genotype–phenotype dichotomy. This is because the target of mutation is the gene pool belonging to the entire population, while the individual, a collective unit of particular genes, is the phenotype exposed to selection. For a genome with more than a hundred million nucleotides, all mutants are unique. They can no longer be described, as was the case for the molecular quasi-species distributions and for populations of viruses, by deterministic population numbers. For such a large genome, there must be many neutral mutations, that is, changes with no accompanying selective advantage or disadvantage. It cannot be predicted which of these will grow up by genetic drift and which will die out. New evolutionary routes are opened by additional mechanisms of genetic change: over and above point mutations and the insertion and deletion of oligomers, we find rearrangements at the junctions of the recombined segments of sequence, or the insertion of doubled or inverted gene fragments. In this way, the horizontal spreading of mutants throughout a population is rapid. The processes that lead to a horizontal spreading of genetic information that short-cuts the vertical mechanism of selective evaluation have collectively been termed 'molecular drive'.[35] It was this that helped the eucyte out of the evolutionary cul-de-sac into which the later replicators, with their of necessity drastically reduced error rate, were predestined to get stuck. The immense learning capacity of the immune system, which recognizes an enormous variety of molecular structures foreign to the organism,[36] is based largely upon the opportunities for variation offered by the recombinative exchange of hereditary matter.

The evolution of higher, multicellular life forms had to await the perfection of the means of recombination. So, right up to the late pre-Cambrian period (around three thousand million years after life began), we find only unicellular organisms. After this point, some five hundred to one thousand million years ago, there suddenly commenced an explosive development that resulted in a shower of miraculous products of evolution. Multicellular organisms demanded new paths of self-organization. It is only the *regulation* of self-organization, and not self-organization in itself, that is programmed into the genome of the cell. This we can illustrate with an example. *Homo sapiens* has more nerve cells in his central nervous system than information symbols in his genome. This means that the contacts that connect these millions of cells cannot be pre-programmed in detail. The only thing that can be genetically predetermined, apart from a few specialized functions, is the way in which the organ is built up, for cell differentiation and morphogenesis are also self-organization processes;[37] they take place on a higher, cellular level, but they are still directed by molecular events. The classic example of a highly differentiated organ is the central nervous system of the human, with its many thousand million nerve cells communicating with each other by way of 1000 to 10 000 contacts each.[38]

10. Unceasing creation

Self-awareness, then, was simply a function of matter organized into life; a function that in higher manifestations turned upon the very matter that bore it and became an effort to explore and explain the phenomenon it displayed — a hopeful–hopeless effort of life to achieve self-knowledge, Nature turned in upon herself — a project doomed to ultimate failure, since Nature cannot be resolved in knowledge, nor can life, in the last analysis, eavesdrop on itself.

The origin of life cannot simply be defined as the transition from inanimate to animate matter. For one thing, the transition as such cannot be pinpointed, as it is a gradual one. For another, it does not conclude the evolution of life; on the contrary, it sets off an uninterrupted chain of developments whose complexity vastly exceeds that of the first steps toward life. Indeed, the earliest autonomous organisms are far more distant from mankind than they are from 'that Nature that did not even deserve the name dead, because it was inorganic'. This is clearly reflected in the temporal progress of evolution. Cellular life appeared on our planet, after it had cooled down enough to allow chemical self-organization, within less than one thousand million years. It is probable that the greatest part of this period was needed for the necessary molecules for the processes of life to accumulate in sufficient concentrations to encounter one another frequently and to organise the first 'social animal', the cell. After this, life lingered at the unicellular level for some three thousand million years. Naturally, this phase is subdivided into many individual steps. The cells had to develop properties that would later allow the construction of cellular social patterns, for which the amoeba today provides a classical example. The path led from molecules to integrated systems of molecules, from the single cell to the system of interacting cells, from the agglomeration of cells to the integrated cellular society, from the organ to the higher organism. It took the best part of another thousand million years before human beings came along. Before us still lies the evolution from man to mankind, from single humans with various degrees of social organization to a true human society.

Life has become an inseparable part of the environment. Species are totally dependent upon one another. We speak of ecological balance, although there can never be an equilibrium in this perpetually changing scene. The innumerable interconnections mean that perturbations can have far-reaching and usually incalculable consequences. The selection value of every species is a complicated, non-linear function of many variables. A dogmatization of the idea of selection and its projection into the social realm of organisms could lead to terrible consequences.

It is clear that the solution of the problem of life — even regarded as an abstract problem of a theoretical nature — cannot be sought in a 'world formula'. Ludwig Wittgenstein captured accurately the situation of modern biology when he wrote in his *Tractatus Logico-Philosophicus*: 'The solution of the problem of life is seen in the vanishing of this problem'.

The process of creation is by no means at an end, although no-one can predict what is to come, even within intervals of time that are negligibly short in comparison with the phase of genetic evolution. Today we are in a position to intervene in, and repair, the genetic process (see Vignette 5). A creative intervention, however, would demand knowledge that we do not (yet?) possess. But evolutionary progress in the near future will hardly be on the genetic level. The activation of the human mind has greatly speeded up the roundabout of development. Almost everything that happens in the foreseeable future will proceed from mankind. Now as always, the motto of evolution is: survival. We will only be able to rise to this challenge by the mobilization of our mind, the ethical component of which is, however, unable to keep up with the breakneck pace of development in science and technology. Here, too, there is no 'world formula' to help us: step by step, we shall have to wrestle for the solutions ourselves. Man is still a relative newcomer to the planet Earth, and the creation of humanity has only just begun.

Part II
Vignettes from molecular biology

1. Sequence comparison (statistical geometry)[4,5,39]

Sequence determination means the chemical identification of the monomeric units at each position in the macromolecular chain of a nucleic acid or a protein. The expression thus implies primarily a chemical analysis. *Sequence comparison* starts out with a set of molecules whose sequences have already been determined, and its aim is to investigate the relationships between the information contained in the different chains. It is thus more of a mathematical analysis. In the following explanation of sequence comparison, we shall concentrate upon nucleic acids, but the same ideas apply also to proteins.

We begin by placing the sequences alongside each other, using common features to ensure that corresponding positions lie side by side. This procedure, called *alignment* of the sequences, is shown in the following example of four sequences, in this case binary sequences (consisting of two classes of symbol, R for purine and Y for pyrimidine).

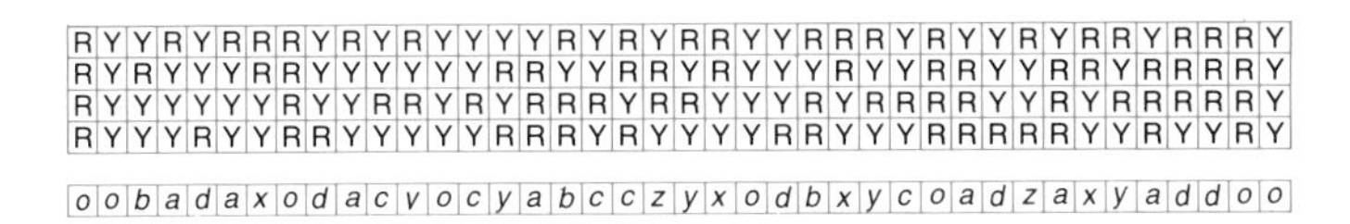

The sequences can contain insertions (extra nucleotides) or deletions (missing nucleotides). For this reason, they are not always of equal length. In such cases, the alignment involves examination of the homologous regions (or, better, regions showing strong similarity) in order to find out where the insertions or deletions are. Alignment is best carried out with the help of several sequences.

Let us see how this procedure can be used to reconstruct the information in a common precursor, with the help of a linguistic example. We consider ten sequences of letters. Each sequence has arisen by random alteration of the single letters of a common precursor. The original meaning cannot be deciphered by examining any one of the ten sequences. However, if we write them on pieces of transparent foil and superimpose them correctly, the original message can be read off at once. For nucleic acids, we call the sequence that arises from this procedure of alignment and superposition the *consensus sequence*.

THY ONIONS OR SPECTRA
TERROR GIV FOXRABIES
TEA OR WINE ORSPECIER
SHERRY GIN O TREMENS
PEAS ICED IF SHELLED
THE KING NOT ELECTED
SHE REAGAN UFO PICIES
TREMORS IN F. REGION
AHA QUIET GO APACHES
TREVOR!GINA FAST KISS

Any pair of sequences can be characterized by a distance that is defined by the number of differently occupied positions. Since different positions in a real nucleic-acid chain will have different abilities to tolerate mutation, it is sensible to classify them in this respect. In the example showing the alignment of four sequences, this would be as follows:

(1) positions at which all sequences are identical — type *o*.

(2) those at which one sequence differs from the others — types *a*, *b* , *c*, and *d*.

(3) those at which two sequences are of one class and two are of the other — types *x*, *y*, and *z*.

This classification is indicated in the first diagram by the row of letters under the four sequences.

The relationships can be represented in the following geometrical figure. The four outermost points A, B, C, and D stand for the four sequences and the number of mutations separating them is in each case given by the total length of the lines connecting them, while each line is equal in length to the number of positions in its respective class above. Thus, in this example, $a = 7$, $b = 3$, $c = 5$, and $d = 6$.

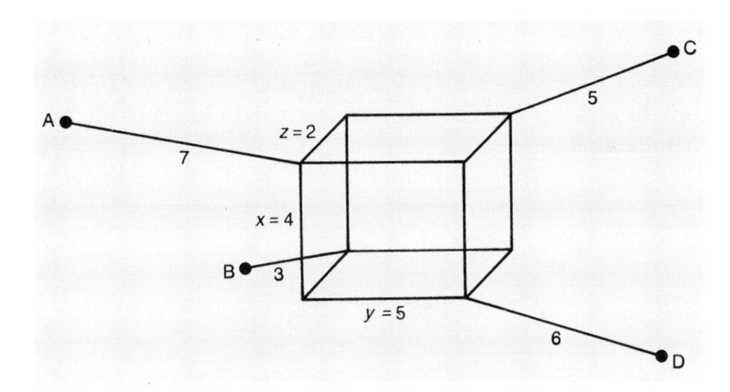

If instead of binary sequences we were dealing with quaternary sequences (such as the nucleic acids, with four different symbol types) then there would be other classes of distance and the figure would be correspondingly more complex (see Vignette 12, where the idea of sequence space is explained). Using the same ideas, we can correlate any number of sequences, obtaining a progressively more complex geometry as the number of sequences compared rises. To represent this geometry, we would need a higher-order diagram in sequence space. A tree would then be a mere first approximation to this diagram.

The value of this method, which is tantamount to a comparison of distance categories, becomes clearer when it is used to compare a larger number of different sequences. One can do this with the help of averaging, so the method is termed statistical geometry. Naturally, all possible combinations of four nucleotides must be considered; however, for forty sequences this means nearly 100 000 combinations, so the task can only be mastered with the help of a computer.

Statistical geometry allows the topology of the distribution of sequences to be assigned objectively. Let us consider the average values of $\frac{1}{4}(a + b + c + d)$ and those of x, y, and z. We denote the longest of the three dimensions of the box in the figure as l, the middle one as m, and the shortest as s. We then apply the rule:

> If $(l + s)$ is roughly equal to $2m$ and this value is small compared with $\frac{1}{4}(a + b + c + d)$, then we are dealing with a bundle-like topology. For an ideal bundle, l, m, and s are all equal to zero.

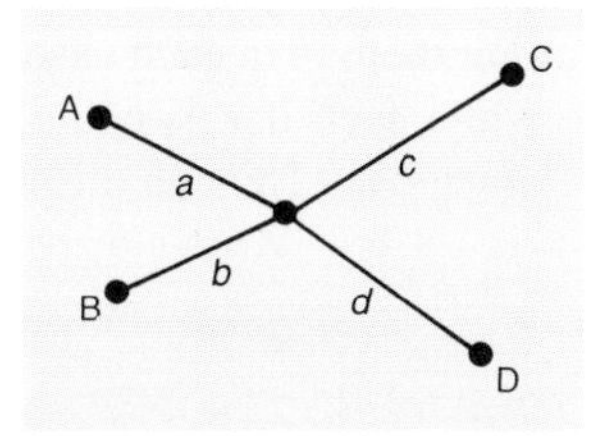

> If l turns out to be much greater than m and s, then we have a tree-like topology. In an ideal case, m and s are zero and l is about the same size as $\frac{1}{4}(a + b + c + d)$.

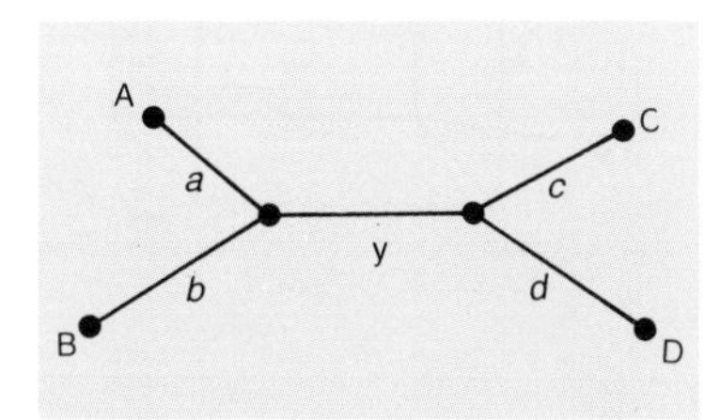

If, finally, all the measurements are about the same, then sequences in the distribution are interrelated in a network-like manner. This is the ultimate state of blurring of the information, irrespective of the kind of topology it comes from. In this case, any resemblance among the sequences is purely coincidental.

If averaging is carried out over very many combinations, then statistical geometry gives reliable information. This applies especially in the case of quaternary sequences. In contrast, the mere addition of distances fails to yield any reliable information about the degree of randomization, since some points of resemblance can be due to functional constraints. In such cases, short distances do not necessarily indicate a low degree of randomization. For homogeneous mutation rates, the different distance classes change in a quantitatively correlated manner. This quantitative correlation, and deviations from it, are important in more advanced applications of this theory, because they allow the analysis of kinship to be refined and used for comparisons extending over a greater distance.

2. Sequence comparison (examples)

Some sequences are very variable, while others have remained almost unchanged during evolution. Among the least variable sequences are those of the functional units of the cell's machinery for reproduction and translation. For our first example, we take the case of a phylogenetic divergence lying so far back in the past that tree-like branching can scarcely be distinguished from a bundle structure. The four sequences — they are those of a special type of ribonucleic acid, called 5S-rRNA — have now gone their separate ways for nearly three thousand million years. A and C are sequences from eubacteria: A is *Bacillus pasteurii* and C is *Anacystis nidulans*, a cyanobacterium (a blue-green alga capable of photosynthesis). Sequences B and D are from archaebacteria, from two different lines that parted company at an early stage: B is *Halobacterium salinarium*, and D is *Methanococcus vannielli*. Their common branching-point lies somewhere within the cuboid whose dimensions (l, m, and s; see Vignette 1) reflect the amount of statistical fluctuation, which can be termed 'sequence noise'.[4,5]

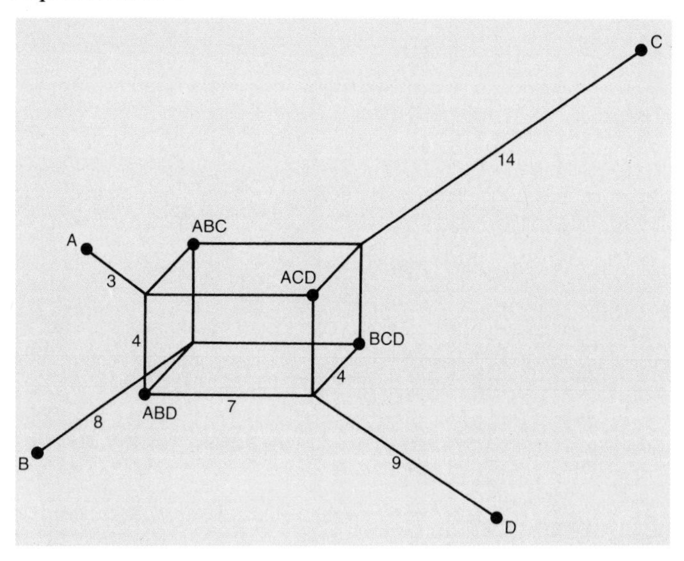

Particularly well conserved are the transfer RNAs[40], the 'adaptors' of the genetic code (Vignette 7). They originated during the pre-cellular phase of evolution, namely, the phase during which the genetic code developed and became established. They are thus among the oldest biological molecules that we know. Before their existence, any genetic code in the modern sense would be hard to imagine. If the sequences of these molecules had not been conserved with unusual accuracy, their original information content would long since have been blurred beyond recognition. In the last hundred million years they have hardly changed at all. For some molecules in this class, *Homo sapiens* and the frog *Xenopus laevi* possess exactly the same sequence. Transfer RNA (tRNA) therefore offers a class of molecules most likely to yield some clues about the initial phase of the origin of genetic information and the beginnings of evolution. The figure shows the divergence of four tRNA molecules from the bacterium *Bacillus subtilis*: those for the adaptors for glycine (anticodon GCC), for alanine (anticodon UGC), for aspartic acid (anticodon GUC) and for valine (anticodon UAC). These are among the amino acids that appear the most frequently in experiments simulating prebiotic synthesis, and they presumably belonged to the first monomeric building-blocks of the proteins. These four 'decoders' thus arose almost simultaneously, more than three thousand million years ago, from a single precursor. Their sequences have since evolved — independently — in a bundle-like way.

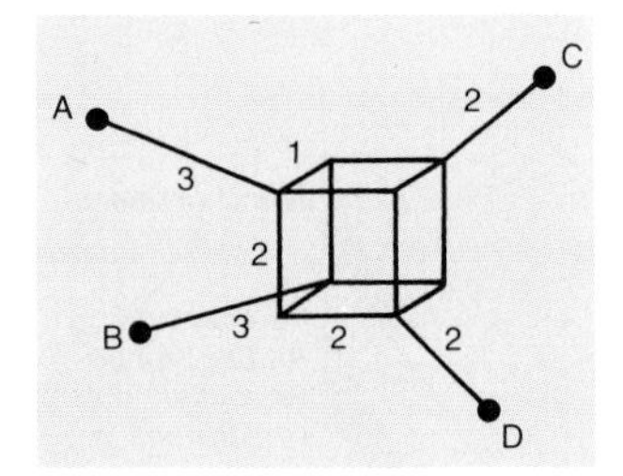

The following example represents unmistakably tree-like divergence ($m = k = 0$). In contrast with the two previous examples, we are dealing here with a highly variable system, one that is still at an active stage of divergence. Only a few per cent of the 990 positions of this molecules have mutated within the short period of observation. The gene in question is that of the influenza virus (A-type), the evolution of which over the last 53 years can today be observed by sequence analysis of preserved samples.[41] Sequence A was isolated in the year 1933, B in 1934, C in 1968, and D in 1985.

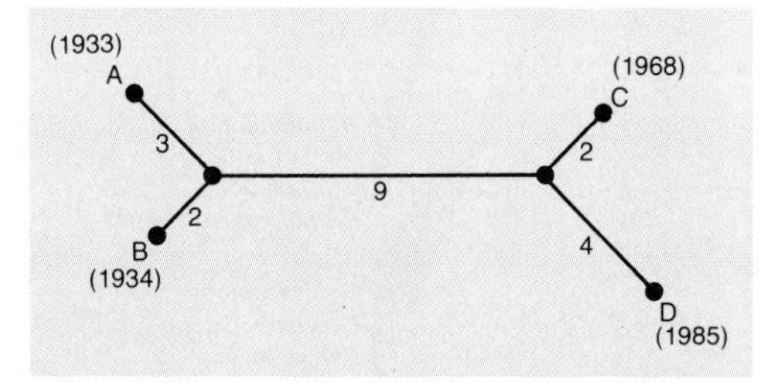

The comparison of sequences reveals the great changeability of the virus, which evolves millions of times more rapidly than the autonomous unicellular organisms considered above. The high rate of evolution is explained by the error-threshold relation in Vignette 10. The variability that results from the high error rate of replication of the virus sets up almost insuperable obstacles for the development of a lasting immunization against the influenza caused by this virus. Similar difficulties have been encountered in the case of the human immunodeficiency virus (HIV), which causes AIDS.

Examples of phylogenetic analysis from sequence data are presented with the help of two dendrograms. The early divergence of the archaebacteria and the eubacteria, and their later dispersion into various species (caused partly by the evolution of their host organisms), were reconstructed from a sequence comparison of 5S RNA and tRNA molecules,[4,5,39,40,42] both functional units of the apparatus of transcription. Although the absolute values of the divergence differ, about the same degree of blurring is found for all the organisms considered. This makes sense in so far as all organisms that exist today are descended from a proto-cell and have therefore incorporated mutations over the same period.

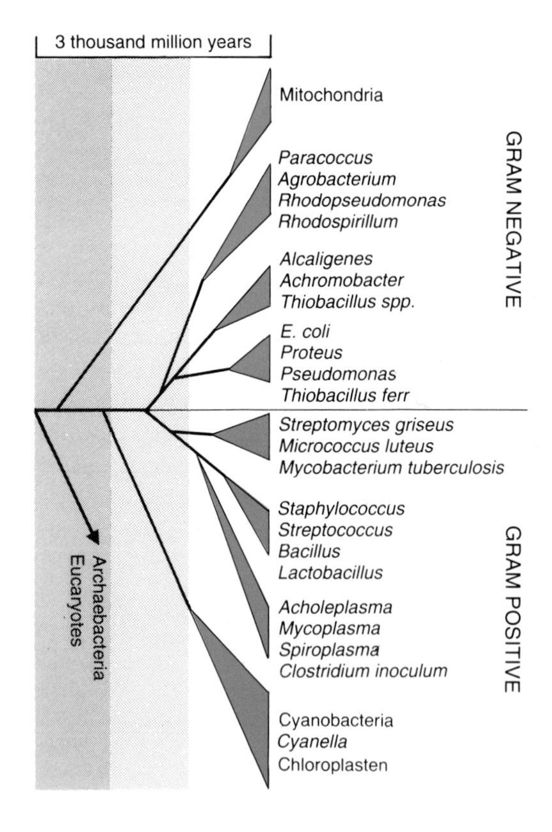

The analyses shown in this vignette have so far been based upon binary sequences: in the analysis of nucleic-acid sequences, only the two base classes (R and Y) were taken into consideration. The next example, in contrast, is based upon the analysis of a protein, in which there are twenty different kinds of monomer to consider. It shows the divergence of the eucaryotes in the last five hundred million years. This was investigated by examination of the sequence of the protein cytochrome c, an enzyme of the respiratory chain.[42]

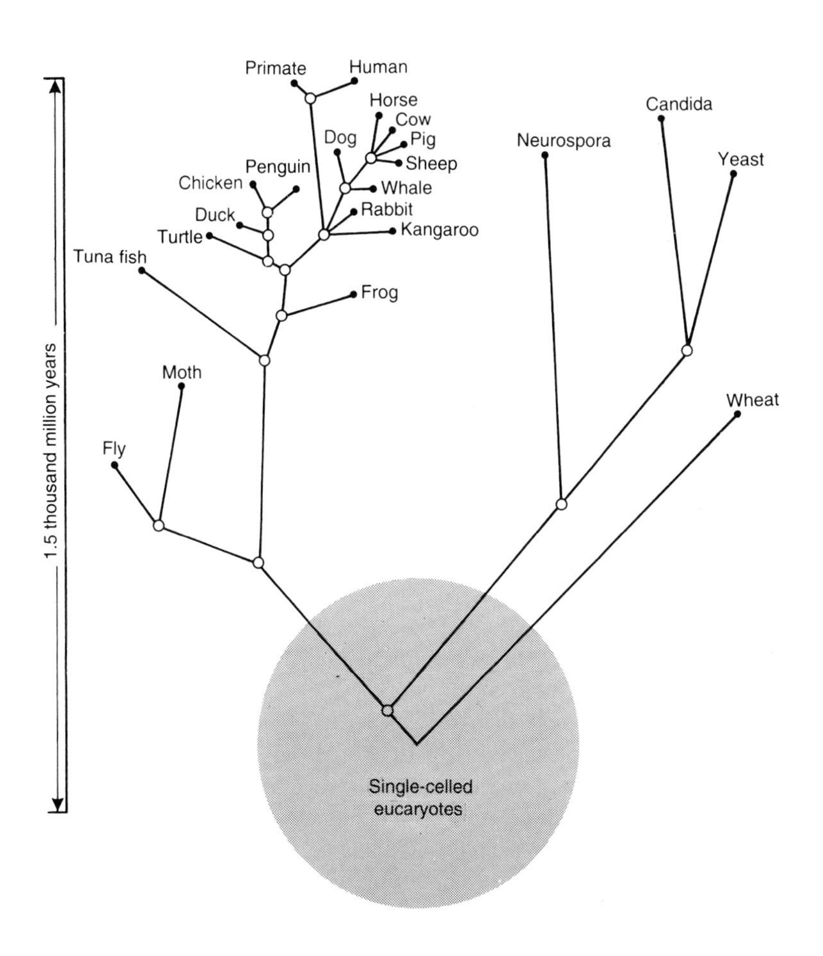

3. How old are the first life forms?

The reply to this question will naturally depend upon our definition of 'life'. If we assume that the origin of life depended *inter alia* upon the ability to store and to pass on information, then the question is as good as answered when we have found out how old the genetic code is (see Vignette 8).

There are about a thousand known tRNA sequences.[40] The job of these molecules is to present each amino acid (protein monomer) for incorporation into the protein when it is called for by the appearance of its code combination in the genetic message. For this reason, they are also called the *adaptors* of the genetic code. For each of fifteen different organisms, including bacteria, algae, plants, and animals, we know the sequences of twenty to forty individual tRNA molecules, the identity of each of which is defined by its anticodon. The different tRNA types arose at about the same time, the period of the origin of the genetic code. They arose from a single distribution of mutants. Since then, they have developed more or less independently of each other.

In addition, there are about twenty-four tRNA adaptor types whose phylogeny has been determined. The sequences of each of these twenty-four given tRNAs have been determined for twenty to forty different organisms. Thanks to these known phylogenies it has been possible to reconstruct twenty-four precursor sequences that refer to the period of the first cellular divergence some three thousand million years ago.

We are therefore in a position to compare three periods: (1) from the origin of the genetic code to present-day organisms, (2) from the very first branching-points (eubacteria, archaebacteria, mitochondria, eucaryote) to present-day organisms, and (3) from the origin of the genetic code to the first branching-points. In all three cases we know the average divergence, that is, the average number of positions at which two sequences differ. The evolutionary tree sketched here is based on these data.

We should, however, note that relative distances do not immediately allow us to deduce relative time intervals. That would only be possible if mutation rates were uniform for all positions of the sequence and had remained constant during the whole of evolution. And they have not: evolution proceeded more rapidly at the start than in later phases. This is because the error rate was very high at the start, because the error-correcting mechanisms of today had not yet evolved, and it decreased later as more and more information accumulated. For this reason, relative distances in the initial period must be taken as upper limits. We thus deduce that the genetic code originated less than one thousand million years before the branching of the archaebacteria and the eubacteria took place. In other words, the genetic code is no older than about four thousand million years. Although the result will not surprise biologists, it is in fact the first experimental determination of the age of the genetic code. There was previously no way of dealing with the objection that the age of the Earth perhaps did not suffice for the evolution of such a complex entity as the proto-cell, with all its elaborate biochemical apparatus.[6]

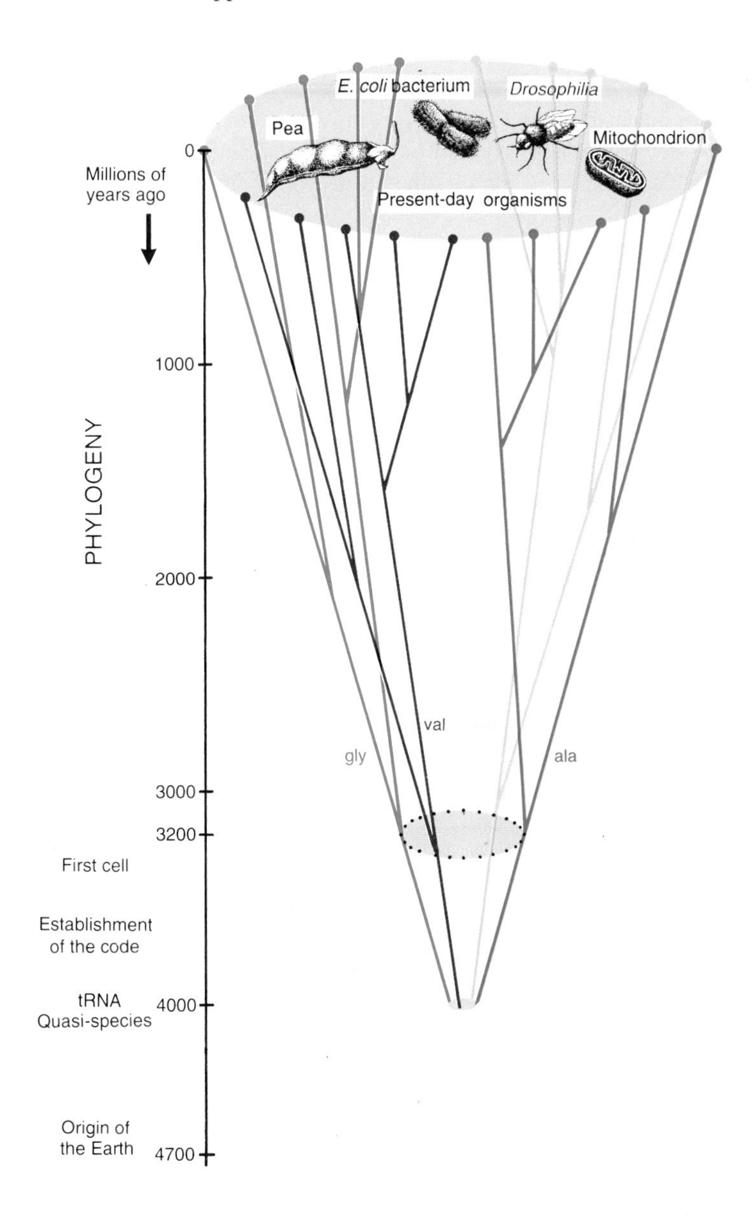

4. Nucleic acids as information stores: the transition from chemistry to biology[31]

The logic of living matter has its roots in physics and chemistry, but very early in its development it adopted its own forms and rules. The nucleic acids lie at the border between chemistry and biology. Their special chemical properties provide the prerequisites for the emergence of living organisms from inert matter.

The monomeric sub-unit of the nucleic acids, the nucleotide, is a defined chemical compound.[43] It is assembled from three different molecular entities: a phosphate ion, a sugar molecule (ribose or deoxyribose), and a cyclic organic compound termed a base:

If the monomers are present in an energy-rich form (for example as triphosphates), then they can unite to give macromolecular chains, even without the catalytic assistance of enzymes.

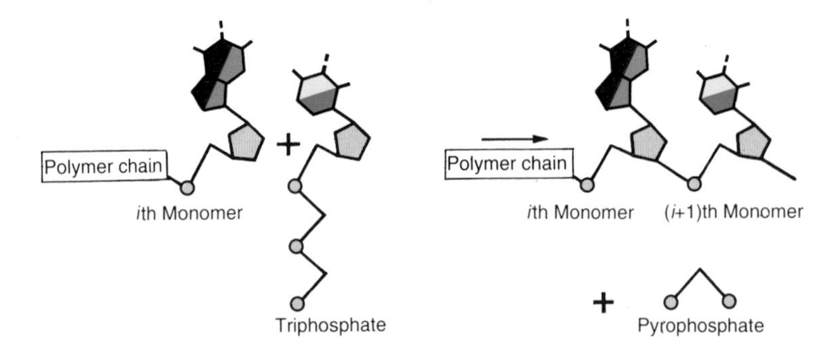

This is still a purely chemical process. But, even at this level, a vital precondition for selection is fulfilled: the growing nucleic-acid chains are metastable. They arise spontaneously from energy-rich monomers and then decompose, in an aqueous environment, into energy-deficient fragments. Growth and decay are thus irreversible processes. Nucleic acids in the presence of water are not in a state of chemical equilibrium. If they were, they would never have been able to develop their characteristic ability to generate and optimize genetic information. As a price for this, organisms have to use a part of their metabolism in the continuous production of new, energy-rich monomers. Indeed, in the course of evolution, the monomers of the nucleic acids (especially adenine triphosphate, ATP) have taken over the role of the cellular fuel that keeps all the cell's biochemical reactions going, and which the higher organisms use to power their muscles.

With the self-replication of nucleic acids, we have arrived at the fundamental step leading from chemistry to biology: the four bases of the nucleic acids begin to play the part of linguistic symbols. The sequence of these symbols can encode a message. There thus arises an attribute that is new to the world of physics and chemistry, one unknown to the science of material interactions, of atoms, molecules, and crystals: that of information.

The definition of information requires, first of all, a restricted set of symbols; secondly, the concatenation of these symbols into chains, or sentences, whose structure is defined by a grammar and whose meaning is realized by semantic agreement; and thirdly — a requirement often tacitly ignored — an apparatus for the reading (and, if necessary, for the translation) of the message contained in the symbol sequence.[7] All of these three requirements for the use of nucleic acids as information stores are fulfilled on the basis of chemistry, as the following facts show.

- The nucleic acids employ four chemically defined compounds: two purines (adenine and guanine, referred to as A and G) and two pyrimidines (thymine and cytosine, referred to as T and C). The base T is used in deoxyribonucleic acid (DNA) only; the sister molecule ribonucleic acid (RNA) uses the slightly simpler base uracil (U) instead. In addition, DNA uses the sugar deoxyribose, while RNA uses the sugar ribose.
- The concatenation of the nucleotides to give a macromolecule results in a message that is metastable and whose existence is therefore not permanent.
- There exist specific hydrogen-bonding patterns between purines and pyrimidines: A interacts preferentially with T (or U) and G with C. Because of this, we speak of *complementary base pairing*. This is the critical factor that ensures both the legibility of the message and the possibility of its indefinite preservation by repeated copying. In addition, the complementarity provides the basis for recognition functions, which are of value in the translation of the genetic message and in regulation mechanisms.

The two base pairs are similar in shape but differ in their strength of adhesion: the AU pair, held together by only two hydrogen bonds, is less stable than the GC pair, which is held together by three.

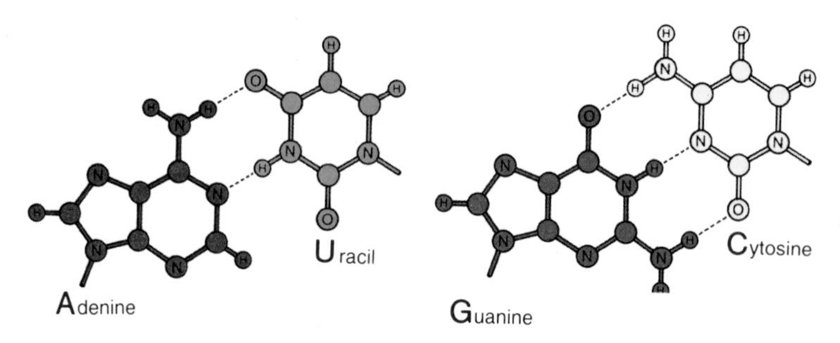

If we use a simplified notation to express the symbols, thus

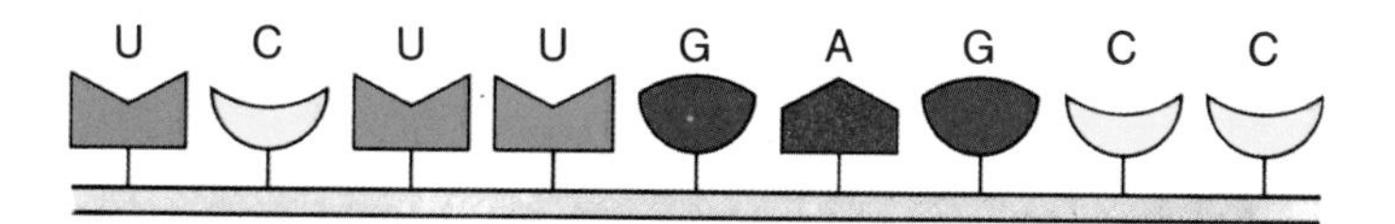

then we can show the replication process heuristically in the following way:

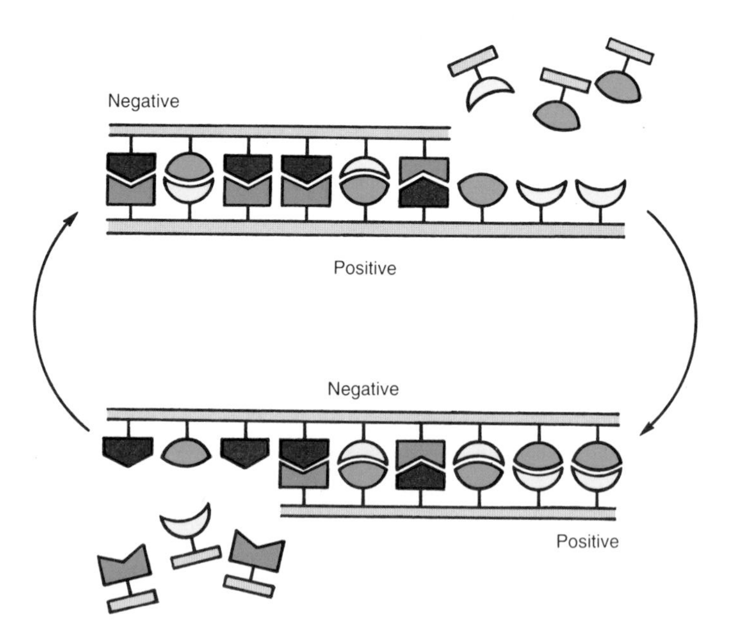

The copying of a chain proceeds, as in photography, by way of the 'negative' as an intermediate product.

A final important point is that replication does not take place with complete accuracy. The complementary interaction has only a limited strength, so it can be disturbed both by the thermal movement of molecules and by chemical interference. In this way, copying errors are made, leading to mutants, which provide a reservoir of alterations that can be made use of in evolution. Self-replication, mutability, and the metabolism made necessary by the metastability of the molecular chains — all of them features of living organisms — co-star inseparably with the nucleic acids on the screen of evolution. In addition, the properties of the information store are sufficient to initiate an evolutionary process from which meaningful genetic information ultimately can emerge by a process of evaluation. The exciting history of the discoveries that have led to our present-day insight into this process of molecular evolution is sketched in Note 44.

5. Structural forms of the nucleic acids[45]

The nucleic acids occur primarily in two chemical forms, DNA and RNA. DNA is chemically more inert than RNA, and in aqueous solution it decays (is hydrolysed) less rapidly than its sister molecule. Furthermore, a single strand of DNA tends to unite with a complementary strand thereby forming a stable double helix. In this way, DNA can attain a life-span that is comparable with that of its organism. Whenever it is necessary to degrade a DNA molecule, special enzymes are employed.

The figure below shows the form of DNA whose structure was first elucidated, by James D. Watson and Francis H.C. Crick. This is the form most frequently used by organisms to store their genetic information.[46]

Not only does RNA take on the role of a messenger (mRNA), but it is also entrusted with a variety of functions in the cell's machinery of reproduction and translation. Its sugar–phosphate bond is chemically more vulnerable than that of DNA, which gives RNA a considerably shorter life-span. However, it forms somewhat stronger base pairs, which leads to pairing in short, complementary regions within a single strand; this occurs with such avidity that it hinders the recognition of entire complementary strands, if these are present, and thus it usually prevents the formation of double-stranded RNA molecules analogous to DNA. However, the incomplete self-pairing of single-stranded RNA molecules that are internally not fully self-complementary leaves many unpaired bases exposed, and these are particularly susceptible to chemical attack, increasing the inherent vulnerability of RNA to hydrolysis.

RNA fulfils its messenger function by conveying the information — copied by an enzyme from the DNA sequence of the gene into the sequence of new messenger RNA — from the gene to the site of protein synthesis in the cell. Here it acts as the template that directs the synthesis of the protein, which then can carry out its function as an enzyme or as a regulatory factor in the cell. When it has done its job, the mRNA molecule is degraded by enzymes. Control of the synthesis and the degradation of mRNA is one of the ways in which the synthesis of proteins is regulated in the cell.

In evolution, RNA came before DNA. It follows from this that most of the functional nucleic acids (those not concerned with conveying information) are RNA molecules that became fixed in their roles at an early stage of evolution. These include the ribosomal ribonucleic acids (rRNA), which are structural and functional elements of the ribosome, the cell's 'protein synthesis factory', and it includes the transfer ribonucleic acids (tRNA), which act as adaptors between the 'words' to be decoded and the amino acids that these 'mean'. These nucleic acids are in general single-stranded and have a well-defined spatial structure. As

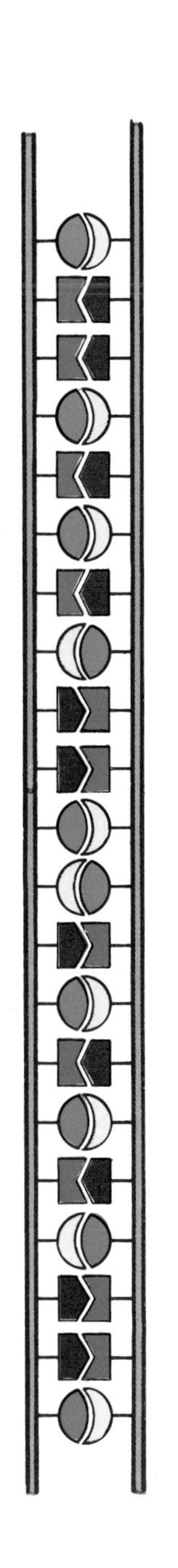

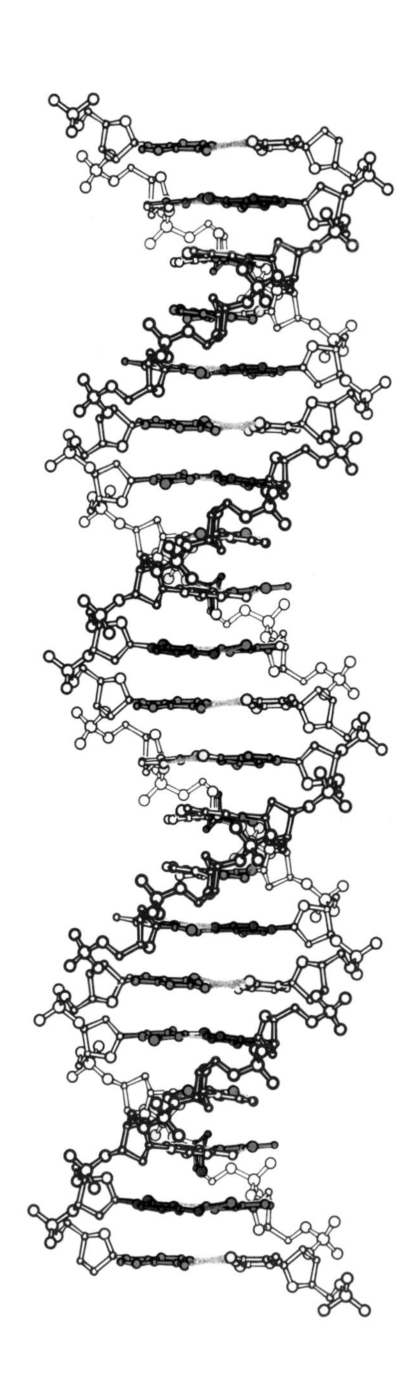

an example, we show the secondary (base-pairing) and tertiary (spatial) structures of the tRNA that recognizes the codon UUC. The base-pairing pattern shows a characteristic 'clover-leaf' pattern of paired bases, adopted by all species of tRNA.

The functionally important structure is the folding in three dimensions. It should be noticed that the anticodon (positions 34, 35, and 36) and the amino acid bound in position 76 are prominently exposed, and this is of course to be expected in view of their function, since they must be available for interaction with other molecular species.[47]

Further important examples of single-stranded RNA molecules can be found in the viruses.

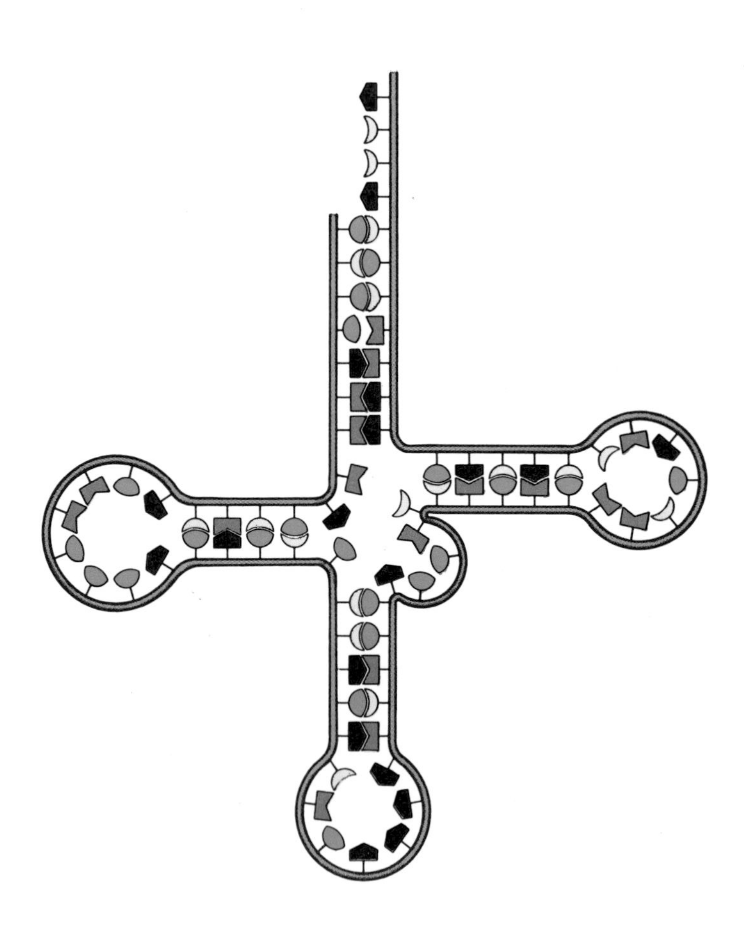

3' end
Site where amino acids are bound
5' end
Anticodon bases

6. The proteins: molecular purveyors of cellular function[31]

The proteins are the molecular executive of living systems. As catalysts (enzymes) and as regulators (promotors, repressors etc.) they steer the entire chemistry of the cell. They also act as receptors and as material for the construction and reinforcement of cellular structures; these functions they share with the lipids (fat-like substances), the polysaccharides (long-chained sugar polymers), the lipoproteins (molecules with a protein and a lipid component), and the glycoproteins (molecules with a protein and a polysaccharide component). Since proteins, unlike nucleic acids, cannot be 'read', their synthesis must be programmed, or *instructed*, by nucleic acids (see Vignette 7). However, thanks largely to the wide chemical variety of their various monomers, the proteins offer a vast range of adaptable functional structures. The monomers of the proteins are the amino acids. Like the nucleic acids, these can be condensed to macromolecular structures.

Two amino acids can combine and expel a water molecule to give a dipeptide. The individual nature of the amino acids is retained, since the reaction, or *condensation*, does not affect their side chains (in the figure, R_1 and R_2).

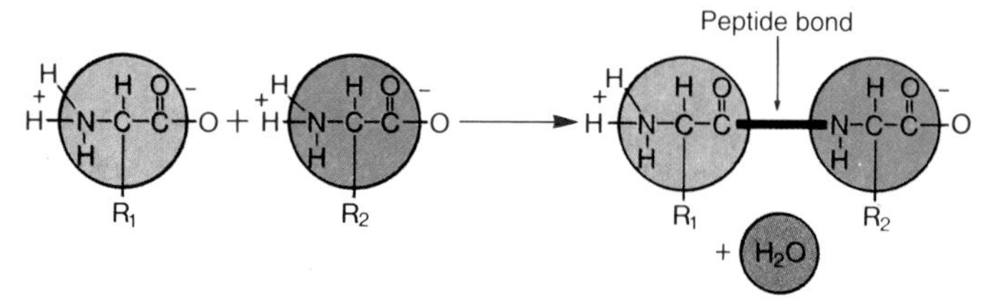

Repetition of this process leads to the linear polypeptide chain, termed the *sequence* or *primary structure* of the protein. As an example, we show the sequence of cytochrome c, an enzyme from the respiratory chain.

The full and abbreviated names of the amino acids are:

alanine (Ala, A)	leucine (Leu, L)
arginine (Arg, R)	lysine (Lys, K)
asparagine (Asn, N)	methionine (Met, M)
aspartic acid (Asp, D)	phenylalanine (Phe, F)
cysteine (Cys, C)	proline (Pro, P)
glutamic acid (Glu, E)	serine (Ser, S)
glutamine (Gln, Q)	threonine (Thr, T)
glycine (Gly, G)	tryptophan (Trp, W)
histidine (His, H)	tyrosine (Tyr, Y)
isoleucine (Ile, I)	valine (Val, V)

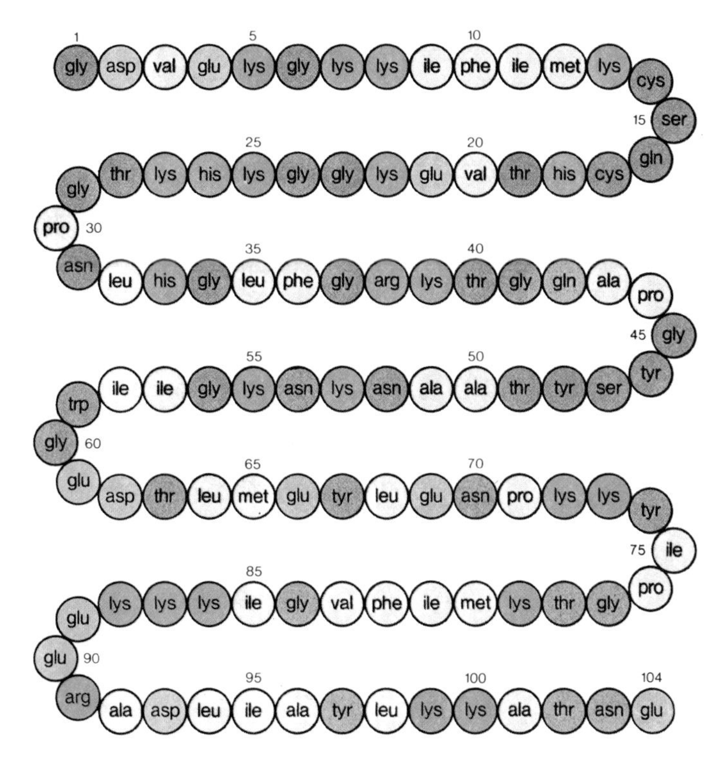

The colours are used to distinguish four kinds of side chain R:

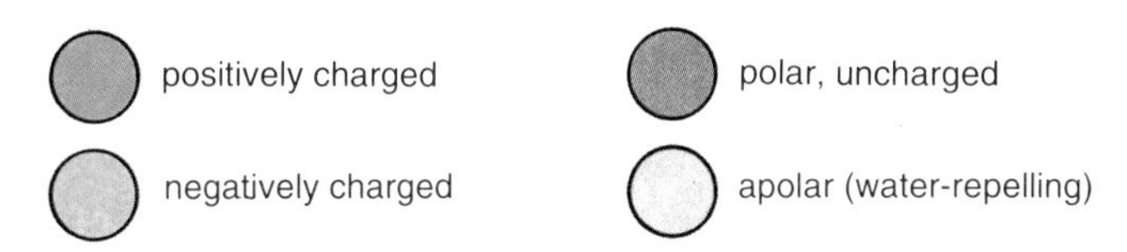

The linear polypeptide chain is folded in a specific way. We distinguish between its *secondary structure* and its *tertiary structure*. The secondary structure arises from the hydrogen bonds between the CO and the NH groups of the peptide chain. A particularly common structure is the α-helix, in which the polypeptide chain winds up to give a spiral-staircase structure with hydrogen bonds joining each CO group to the fourth nearest NH group in the sequence. Another is the β-sheet, a flattish pleated structure in which parallel or antiparallel strands of the polypeptide chain are linked at regular intervals by hydrogen bonds between NH and CO groups. In our example of cytochrome *c*, the regions 1 to 12 and 89 to 101 possess the helical structure.

The tertiary, or three-dimensional, structure of cytochrome c is shown below (after Dickerson[48]). The tertiary structures of proteins are produced, and stabilised, by interactions between the side chains. This structure leads to the formation of an active centre. In cytochrome *c*, the active site is made up of a porphyrin ring containing an iron atom at its centre. The state of oxidation of the iron (Fe^{2+} or Fe^{3+}) determines its ability to act as a donor or an acceptor of electrons.

In consequence of the exact folding of the polypeptide chain, the active centre is *substrate-specific*. The catalytic activity of the iron complex is enhanced by the side chains of the amino acids. A change in the spatial structure of the protein would result in a dramatic decrease in its catalytic activity. In the illustration, only the positions of the central atoms of the polypeptide backbone are shown (in colour, corresponding to the above scheme), while the side chains are shown in grey. With the help of X-ray diffraction,[49] the position of every atom in this giant molecule can be observed with a resolution of 2 Ångström units.

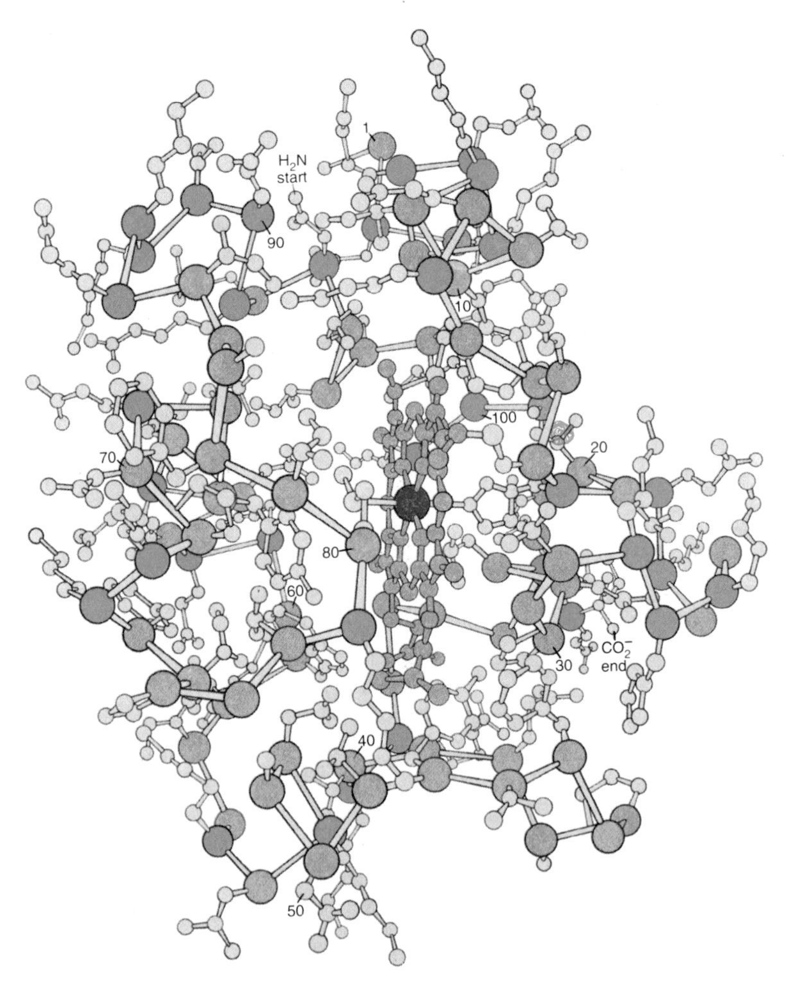

7. Instruction: copying, reading, and translation[31]

All functions involving the transfer of genetic information are based upon the complementarity of the nucleic-acid bases. This clearly applies to the replication of the DNA in the genome of the cell, which preserves the genetic information over indefinite periods, but it is just as true of the rewriting (transcription) of the information into the shorter-lived mRNA, and of the synthesis of proteins on the mRNA template. Each coding triplet (*codon*) in the genetic message is associated with an adaptor molecule, tRNA: the tRNA has in the middle of its sequence a base triplet (*anticodon*) that is complementary to the codon, and bears at one of its ends the correct amino acid (see Vignette 8).

The copying of DNA, the transcription of the message in DNA to mRNA, and the synthesis of proteins on the mRNA template are summarized in the accompanying scheme, which, complex as it may appear, is in fact a highly simplified representation of the processes that really occur.

The reproduction of the genetic message is catalysed by the enzyme DNA polymerase (1), its transcription by DNA-dependent RNA polymerase (2), and its translation by the ribosome. The coupling of the amino acids to their respective adaptors is carried out by the aminoacyl synthetase enzymes (3). The complete mechanism includes additional proteins with both regulatory and catalytic function, and it is assisted by chemical energy supplied by the energy-rich triphosphates of the cell. Taken as a whole, the process is ultimately a huge autocatalytic feedback loop, since the entire machinery is encoded in the DNA and can therefore be said to 'produce itself'.

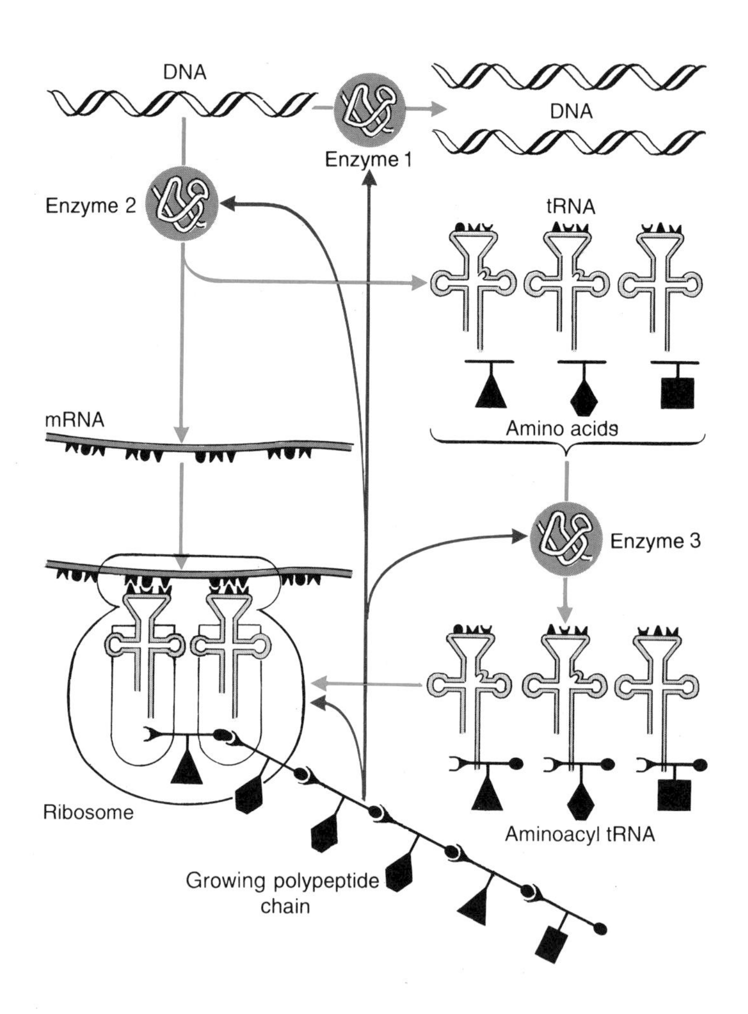
DNA
Enzyme 1
DNA
Enzyme 2
tRNA
mRNA
Amino acids
Enzyme 3
Ribosome
Aminoacyl tRNA
Growing polypeptide chain

8. The genetic code[31]

There are four nucleic acid monomers and therefore $4^3 = 64$ possible triplet combinations of these. The genetic code[50] is a directory that states the assignment of the amino acids to their respective triplets. Each codon is assigned unambiguously to one amino acid or to the instruction 'terminate synthesis'. Since there are sixty-four codons and only twenty amino acids, nearly all the amino acids are associated with more than one codon. This redundancy is advantageous in a functional language such as the one represented by the proteins. The genetic code is shown in the Figure. The directory is read in the order left side, top, right side; for example, the codon UGG specifies the amino acid tryptophan.

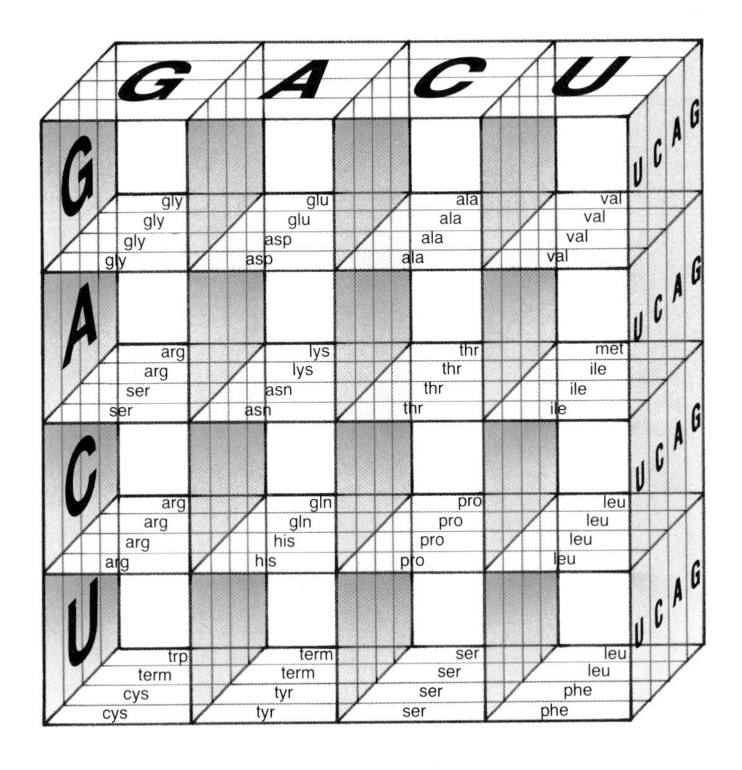

In general, information flows from DNA via RNA to proteins, and these are responsible for all other functions. This direction of information flow was formerly called the 'central dogma' of molecular biology.

There exist, however, certain enzymes that 'copy in reverse' (*reverse transcriptases*). These transcribe information from single-stranded RNA into double-stranded DNA, and they are found in the *retroviruses*[51] (see Vignette 13). In this way, they manage to incorporate their own information into the infected cell, and this is frequently associated with transformation to a tumour cell.

Information residing in protein chains can no longer be translated back into a collinear RNA molecule. There are, however, certain reproduction enzymes that can produce RNA chains *de novo* without the help of a nucleic acid template.[52] The products of the synthesis *de novo* contain short-chained patterns that bear the stamp of the protein's active centre.

These deviations from the general rule are special cases in biology, but are becoming important in evolutionary biotechnology (Vignette 11).

9. Quasi-species: the cloning of mutant distributions

A sequence determination in the laboratory usually requires some 10^{10} RNA molecules. For an influenza virus, this is less than a ten-millionth of a gram. If the result is a clearly defined sequence, then it is easy to conclude that all, or nearly all, of the molecules consist of this same sequence. But such a conclusion would be too hasty, as the linguistic example in Vignette 1 has shown. The only thing we have found unambiguously is the dominant occupation of each position. If at each position only one-hundredth of the sequences deviate from the dominant pattern, then for an RNA chain with ten thousand monomers there is scarcely a chance that even one of the molecules in the sample will correspond to the dominant pattern over its entire length. Charles Weissmann and his colleagues were the first to investigate this, using a distribution of mutants of the bacteriophage Q_β.[14,53] To do this, it is necessary to establish a set of descendants of a single sequence, and then to determine their consensus sequence. The set of descendants of a single mutant is called a clone.

The first step in cloning consists in catching single virus particles, or single DNA or RNA molecules. One starts with a suspension of known concentration, for example 10^8 virus particles per microlitre. This suspension is subjected to serial dilution, for example by diluting it a hundredfold, taking one hundredth of the resulting solution, diluting this again, and so on. After four such steps, instead of 10^8 virus particles per microlitre, we have on average only one. If we take one hundred separate microlitres of such a suspension, then we get a Poisson distribution: roughly one-third of these samples will contain no virus, one-third will contain one virus and one-third will contain more than one virus. If we want to be sure that no sample contains more than one virus particle, we must continue the dilution, for example so as to give an average number of viruses equal to one-tenth per microlitre, so that one-tenth of the one-microlitre samples will contain one virus and the remaining nine-tenths no virus. Finally, the samples are transferred to a culture of bacteria or other host cells, or to an artificial growth medium, and incubated. The virus particles, if present, can then multiply. When they have produced sufficiently populous clones, these are subjected to the usual procedures of sequence determination.

There is one difficulty associated with this experiment.[54] All the mutants that we expect to find are close relatives of the wild type. As soon as they begin to replicate, they produce mutants of themselves, and one of these mutants is the wild type, which, being more efficient, had previously been predominant in the population. If this appears, it is called a revertant, and it now multiplies and displaces the cloned mutant. It is true that the mutant has a head start, since it has multiplied into many copies before the revertant appears. But once the revertant has arisen, the mutant will have to reproduce with a similar efficiency if it is not to be outgrown by the revertant. And, unless it is a genuinely neutral mutant, it will always be outgrown in the end, since the wild type is more efficient. Neutral mutants, if present in the population in sufficiently high numbers, may be seen in sequence analysis when a position is occupied simultaneously by two different nucleotides; such cases are known for certain small, self-replicating RNA molecules.[13]

A number of quasi-species distributions, both artificial and natural (viruses), have been analysed in recent years.[13,14,52,53] The picture found corresponds to that expected from theory. For virologists, the high mutation rate of the RNA viruses came as something of a surprise. The influenza virus evolves at a rate that is more than a million times faster[41] than that of autonomous microorganisms (see Vignette 2). It possesses a genome that consists of eight consecutive RNA sequences, encompassing about fourteen thousand nucleotides in all. A natural population of viruses is a quasi-species whose error rate has been optimized. For an artificial RNA sequence, tailored to suit the reproduction enzyme of the bacteriophage Q_β, the mutant spectrum of the quasi-species has been determined experimentally.[52] Investigation of the viruses of foot-and-mouth disease and of vesicular stomatitis has led to similar results.[53] If the wild-type sequence in the quasi-species is to dominate absolutely, then it must possess a high selective advantage over all mutants. (A case of this kind is discussed in Vignette 10.) This is possibly the case for the virus that causes poliomyelitis, the polio virus.[54] Unlike influenza, the polio virus cannot use rapid evolution to escape from an immunisation programme, and as a result it has been very nearly exterminated in the Western world.

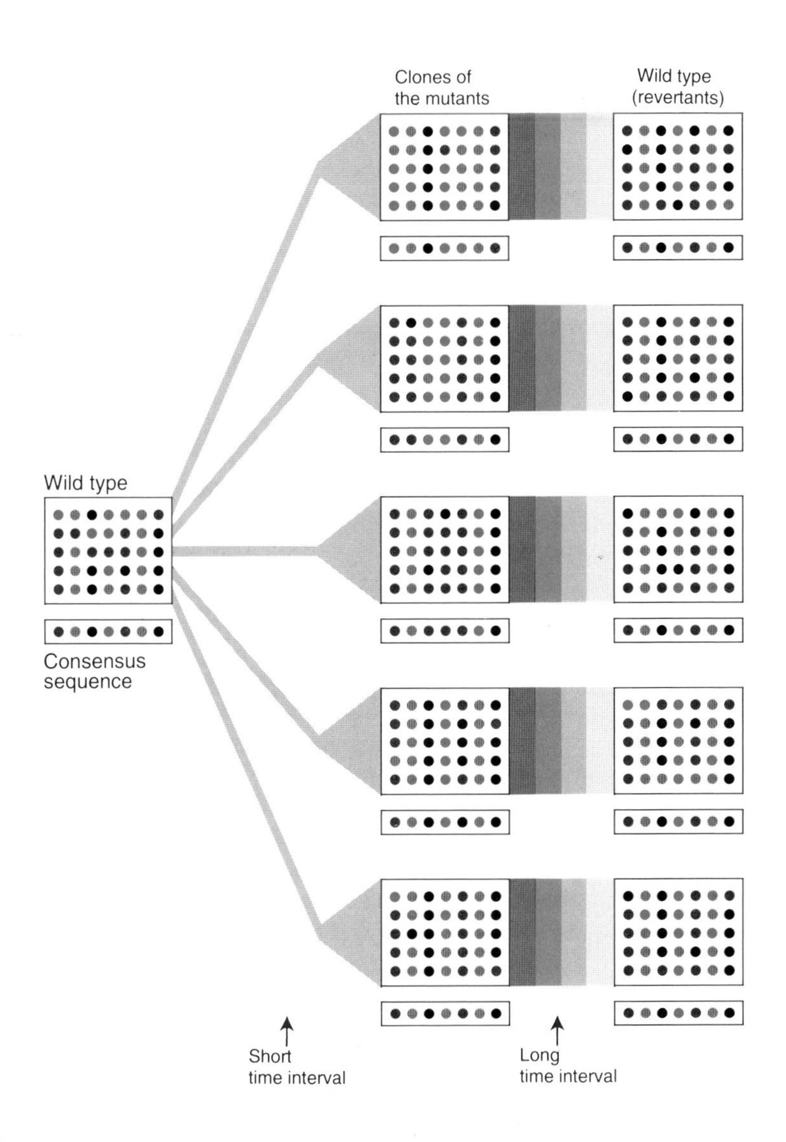
Clones of
the mutants
Wild type
(revertants)
Wild type
Consensus
sequence
Short
time interval
Long
time interval

10. Quasi-species: the structure of mutant distributions[12–15]

In classical population genetics, interest is focused on the wild type, that is, the best-adapted sequence. It has been assumed to be the target of evolution, and to owe its origin to chance, that is, to some random mutation. From cloning experiments (Vignette 9) and from theoretical considerations, we now know that this interpretation is not correct, at any rate not for molecular replicators, viruses, or microorganisms. In these systems, we are generally concerned with a relatively small information content (10^3 to 10^6 nucleotides) and with relatively large populations (10^{10} to 10^{12} or more). Mutants close to the consensus sequence appear reversibly. They are evaluated along with the others in the selection process, and they influence the future development of the distribution, since they can be the forerunners of more advantageous mutants, whose appearance will affect the spectrum of mutants present. The difference between the classical and the molecular picture appears only to be one of scale and number. However, it leads to important qualitative differences, as we shall now see by looking at two examples.

The first is a computer simulation due to P. Schuster and J. Swetina.[55] We start with a best-adapted type, one that is greatly superior to all its mutants: it replicates itself ten times faster than they do. For the sake of simplicity, we assume it to be a sequence of only fifty monomers, which can belong to the symbol classes R or Y. The relative population numbers $\bar{x}_d$ of wild type and mutant are plotted against the absolute error rate (the error rate for the copying of individual symbols), which itself is given by $(1 - \bar{q})$. The index d gives the distance, or number of mutated positions, from the best-adapted type. This distance is called the *Hamming distance*. $\bar{x}_d$ is the number of individuals at a distance d, divided by the total number of all individuals. Thus, the index $d = 0$ refers to the best-adapted species. The respective $\bar{x}$ values (with indices d from 1 to 50) represent the sum of all mutants in the error classes one to fifty: for example, $\bar{x}_1$ includes all the single-error mutants, irrespective of the position at which they deviate from the wild type. The $\bar{x}$ values can vary from zero to one, and their sum is always exactly one. The error rate $(1 - \bar{q})$ gives the probability that a particular position will mutate in copying. An average error rate of 0.01 means that on average every hundredth position is 'wrongly' occupied in the copy. The upper graph shows the population numbers directly, and the lower graph their logarithmic values.

Inspection of the upper graph shows that only at an error rate of zero (on the left) does evolution lead to an all-or-none selection ('survival of the fittest'). As the error rate increases, the population number of the wild type decreases

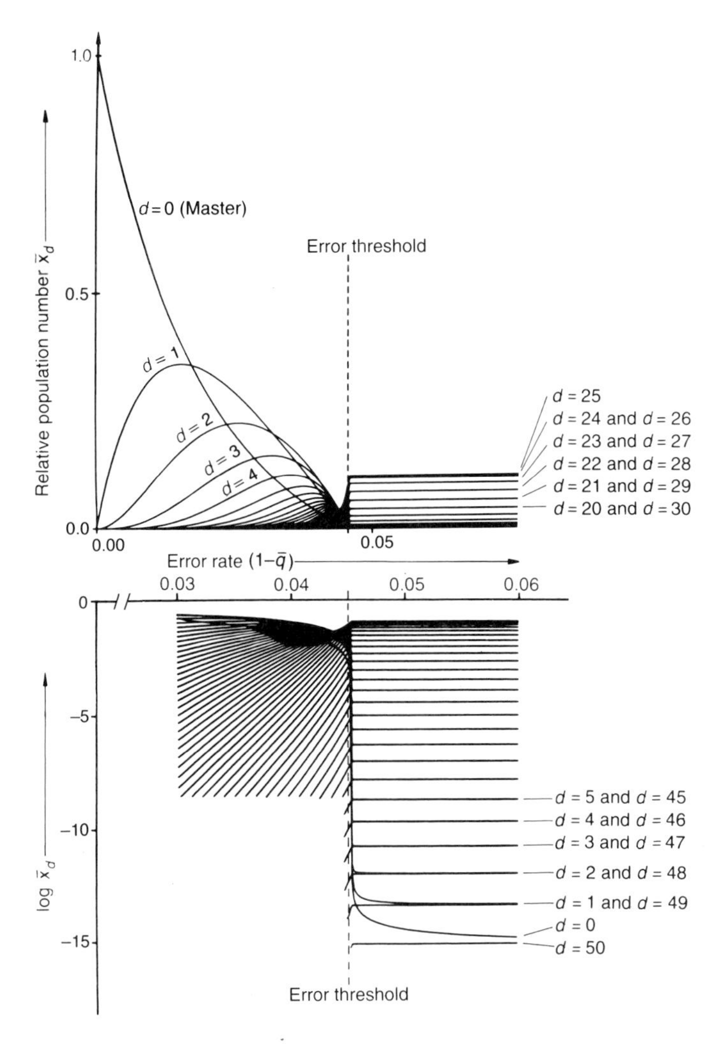

rapidly. Soon there are more mutants than wild-type species, so that the wild type would no longer easily be found in a cloning experiment. Nevertheless, the wild type is still stably selected, since the mutants are distributed, in this case symmetrically, around it. The consensus sequence is identical to that of the wild type. However, when the error threshold is reached, the pattern suddenly changes and the information of the wild type is lost; in this example, the threshold is reached at an error rate of $(1 - \bar{q}) = 0.046$. This can be seen more clearly in the logarithmic plot (lower graph). The proportion of the wild type in the population falls suddenly to about 10^{-15}, which, for a fifty-membered sequence, is the value expected on the basis of random appearance. The wild-type sequence has been relegated to a mere one of 2^{50}, or approximately 10^{15}, possible variants. There are fifty possible single-error mutants, so these are found correspondingly more frequently; the same principle applies all the way up to the twenty-five-error mutants, which as a class are the most strongly represented. Each individual mutant has an expected frequency of 2^{-50} ($\approx 10^{-15}$). The information in the wild type has been lost irretrievably.

The process resembles a phase transition. The error threshold is like a melting-point. The information 'melts' above the error threshold because of an snowballing of accumulated errors. However, just below the error threshold we find the best conditions for evolution. In this region, the wild type is stable as long as no better variant appears in the mutant spectrum. At the same time, the greatest possible number of mutants is produced, some of them — especially in more realistic non-symmetrical distributions — quite a long (information) distance from the wild type. Such a system can adapt very quickly to a change in its environment. Experiments with viruses (Vignette 11) show that natural ensembles of mutants indeed operate just below the error threshold. Systems with an error rate adapted to this appear to possess an evolutionary advantage. The threshold value of the error rate is inversely proportional to the quantity of information (the length of the sequence). The influenza virus possesses about 14 000 nucleotides, and it replicates with an error rate somewhat below 1 in 10 000. The exact level of the threshold depends also upon how much more efficient the wild type is than its mutants. Closely related neutral mutants[19] may be co-selected.

The case that we have considered was a population in which all the mutants were distributed symmetrically around the wild type, and all had the same fitness (expressed by their reproduction rate, which was one-tenth of that of the wild type). Theory shows that the nearly neutral mutants, those whose fitness is nearly as great as that of the wild type, are particularly influential. The second example[56] is intended to illustrate this. In the upper part of the figure, a more complex fitness-landscape is shown. The wild type is in the middle, and the graph shows the efficiency of mutants at increasing Hamming distance (greater number of mutated positions). Stretching over to the left-hand side is a lowland plain similar to the landscape described in the first example, above. The mutants here are at best one-tenth as efficient as the wild type. On the right, the landscape is mountainous: at increasing Hamming distance, the rate of reproduction of the mutants first falls to one-half of that of the wild type and then rises again to attain almost the same value. (A landscape of this kind would be perfectly consonant with the properties of a real virus.) The lower part of the figure shows the population numbers as plotted for the individual mutants. The sequence length is assumed to be 316, and again a binary (R or Y) sequence is assumed. All the population numbers are plotted relative to the wild type ($\bar{x}_i/\bar{x}_0$) and the error rate ($1 - q^2$) is given a value of $\frac{1}{316}$.

The surprising aspect of the computer simulation is not so much the qualitative as the quantitative result. The decrease in the population numbers of individual mutants in the left-hand half of the figure is what we would have expected on the basis of the first example. The error rate chosen causes the

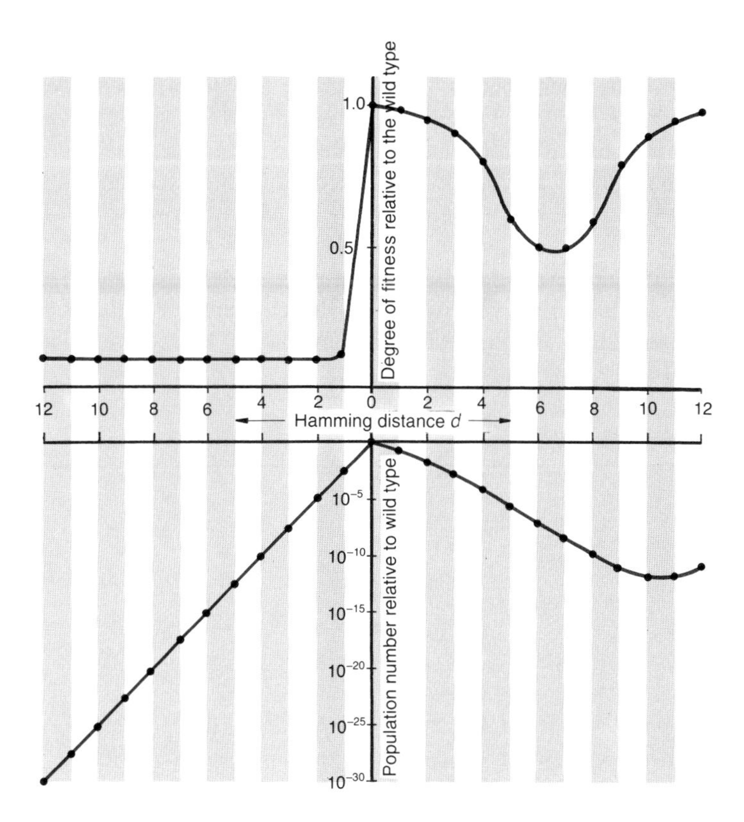

concentration of each individual mutant to decrease by a factor of about 316 for every step away from the wild type. Since there are 316 different single-error mutants, the total number of single-error mutants is equal to the number of error-free wild-type sequences. However, individual twelve-error mutants appear with a frequency of only 10^{-30}. There would be no chance of detecting these in a laboratory experiment, which is normally carried out on only 10^{10} to 10^{12} particles.

The picture is quite different on the right-hand side. The decrease in the number of mutants on the right is much weaker, and at a certain point comes to a halt. The appearance of the twelve-error mutant is predictable and inevitable. The difference in the population numbers of the two mutant distributions is indeed astonishing: in this example, they differ by some twenty orders of magnitude. This is due to amplification based upon a multiplicative relation: the twelve-error mutant arises from an eleven-error mutant, which in turn arises from a ten-error mutant, and so on. Thus the probability of the appearance of a mutant can be influenced; it is much greater along the ridge of a value landscape than in a lowland plain. Since ridges lead to peaks, the evolutionary process can be said to steer itself towards a goal of optimization.

A further difference between this and the classical Darwinian picture should also be mentioned again here. The target of selection is the whole mutant distribution — the quasi-species — and not the individual best-adapted copy. Selection may even favour a copy with a lower individual selection value than the best-adapted one, if the mutants surrounding the former have comparably high selection values while the mutants surrounding the latter have not. In this respect, selection can be characterized by an extremum principle, in many respects analogous to the principle of equilibrium.

These two points reveal sharply the contrast between the classical Darwinian concept of evolution by selection and the concept presented here. The quantitative difference is so great that the new picture leads to completely new qualities.

11. Experiments in evolution

The first 'evolution *in vitro*' experiments were carried out in the 1960s, by S. Spiegelman and his colleagues,[57] who isolated the reproduction system of the RNA bacteriophage Q_β and purified it. On incubating it in solution along with the energy-rich nucleotide triphosphates and the purified, 4200-base-long, single-stranded genome of Q_β, they found that new RNA strands were synthesized; these new strands were just as infectious as their natural precursors. However, if a small portion of these products was transferred to a new incubation mixture and incubated again, and this dilution and incubation were repeated serially, then the RNA molecules grew faster and faster, meanwhile losing their infectiousness. This procedure differs from the serial dilution described in Vignette 9, in that each dilution is made into a new nutrient solution, and growth takes place before the next dilution, so that the loss by dilution is compensated for by the growth of fresh RNA molecules. In this way, the average concentration of RNA molecules is kept approximately constant. However, at the same time, selection pressure is applied so as to favour the sequences that reproduce the most rapidly. If the environment is altered, for example by the addition of an inhibitor, then the evolution of the RNA template automatically becomes programmed to produce a mutant resistant to the inhibitor, and this it duly does.

The idea of this serial dilution with replication is similar to that of a bioreactor or a chemostat, such as have long been used in microbiology. If one wishes to cultivate bacteria in very large quantities (such as ten litres of culture or more), it is preferable to use a continuous-flow reactor instead of a batch process. Either the composition of the nutrient solution is kept constant (chemostat), or else, by regulation of the nutrient input and monitoring of the turbidity on the reactor, the concentration of bacteria is kept constant (turbidostat; the turbidity of the solution is a measure of the number of bacteria present). The figure shows a bioreactor for raising virus cultures, built by Yuzuru Husimi at the author's institute, and it illustrates how the bioreactor works.[58] The turbidostat serves as a reservoir for a bacterial culture of constant population density. This feeds a second flow reactor, the cellstat, in which the viral infection can take place under defined selection conditions, with continuous addition of fresh bacterial cells and removal of cell debris. The graph[59] shows the course of a viral infection. At time zero, *E. coli* cells are infected by the Q_β viruses. After about forty minutes, the first release of newly formed virus particles (about ten thousand per infected cell) is observed, and this recurs at regular intervals. The decrease in the number of virus particles between consecutive 'bursts' is due to the continuous dilution of the viruses by

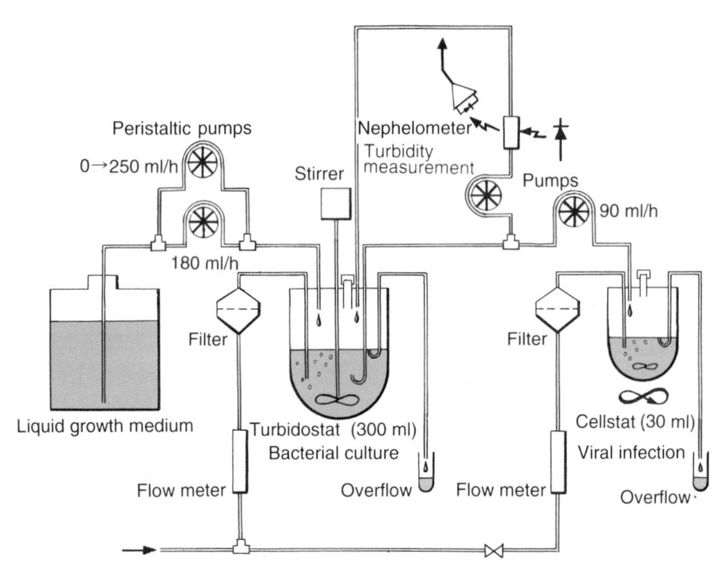

the removal of the solution in the vessel, with the concomitant introduction of fresh *E. coli* cells. The solution ultimately reaches a stationary state, in which the production of viruses is compensated exactly by the dilution. The dilution flux can also be used to introduce other substances, such as antibiotics, in order to exercise selection pressure.

If such experiments are to be carried out on a microlitre scale, it is more appropriate to use the serial transfer method. Here, too, it is essential to arrange for optimal reaction conditions. Insights derived from the quasi-species model can be here applied directly and practically. The evolution process should take place close below the error threshold, where it proceeds with the greatest efficiency. By variation of the rate of mutation, which can be adjusted by changes in the environment, the error threshold can be crossed for brief periods. In this way, an effect is obtained that is reminiscent of annealing, or zone refinement. In zone refinement, a technique used to produce the purest crystals, the melting-point is crossed briefly, and the subsequent recrystallization leads to a curing of lattice defects. In a similar manner, a quasi-species can be optimized by briefly crossing the error threshold. Thus rates of evolution can be increased by many orders of magnitude, in comparison with Spiegelman's experiments, by suitable adjustment of the experimental conditions. This calls for a computer-controlled system that steers the serial transfers. The programme entails the following steps:

- the choice of sample, and thus the definition of the rate of mutation, which depends upon the ratio of the amounts of nucleoside triphosphate in the medium, (G+C)/(A+U);
- the transfer of the RNA template by means of a computer-controlled, automatic pipette into the (still chilled) reaction solution;

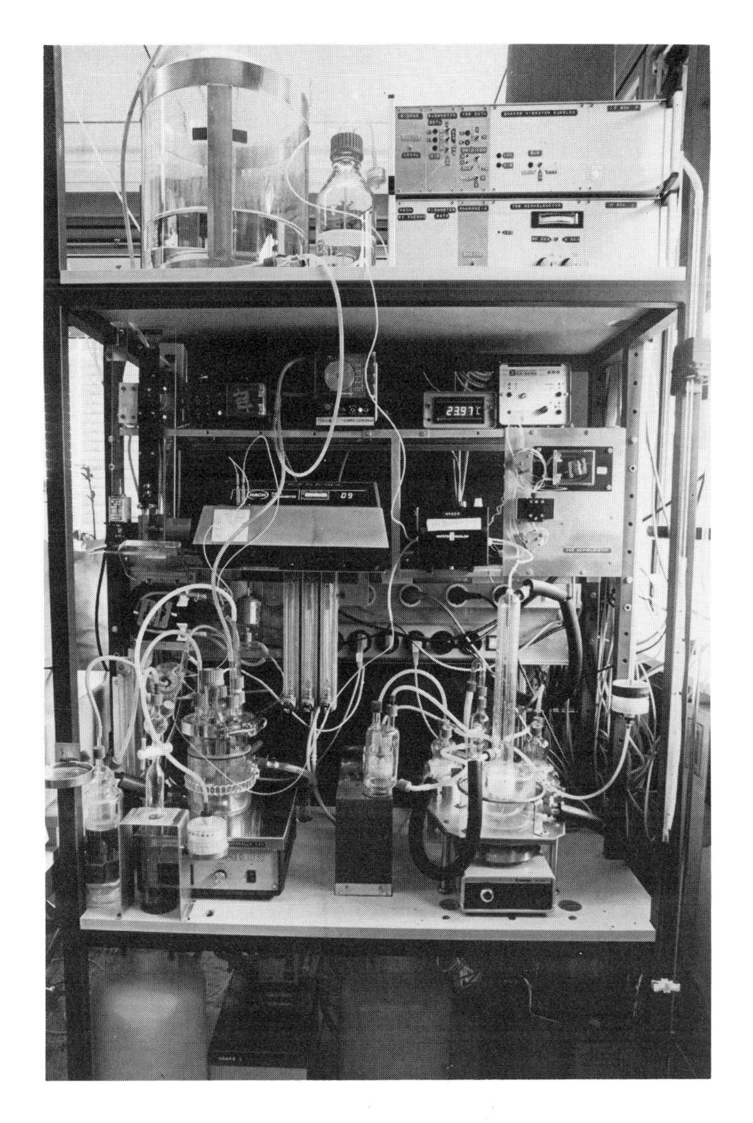

- a temperature jump from 0 to 37 °C or back, within one second, to start or to stop the reaction;
- the control and measurement of the rate of growth by the use of a laser fluorimeter with glass-fibre optics;
- exact adjustment of the physical and chemical conditions, such as the generation of a moisture-saturated atmosphere, the regulation of temperature, and the control of the serial iteration.

From the serial-transfer machine, a whole plant for large-scale biotechnology could be developed. Instead of one or a few samples, a cloning plate could be

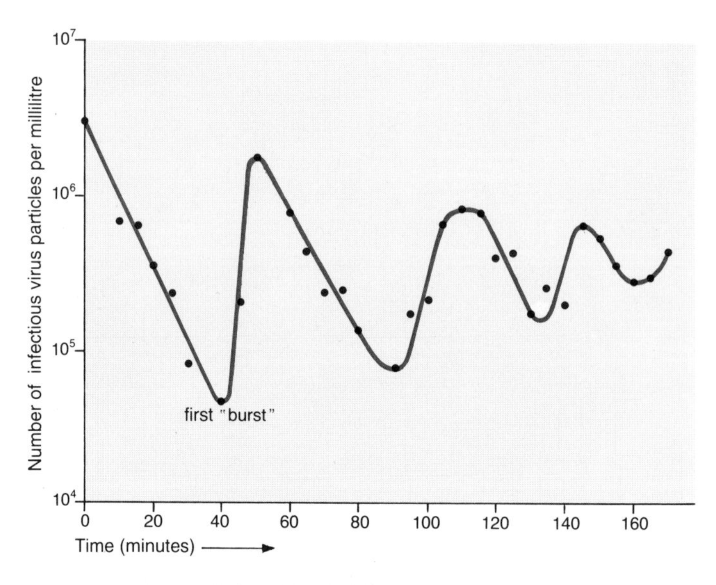

Ⓐ Step motor
Ⓑ Central control unit
Ⓒ Humidity regulations for water-saturated atmosphere at 0 and 37°C
Ⓓ Mechanical micropipette
Ⓔ Serial transfer (0°C)
Ⓕ Replication and fluorescence measurement (37°C)
Ⓖ Sample ejection (0°C)
Ⓗ Sample uptake (below 0°C)

made with a capacity of, say, one thousand samples of a few microlitres each. Naturally, interest would not centre upon molecules with a high rate of replication but upon more specialized properties. Artificial selection would then be realized as follows. By variation of the mutation rate, a mutant spectrum with hierarchically ordered distance relationships would be produced. A multichannel laser fluorimeter would register the efficiency of the reaction products, and a multipipette system, movable in three dimensions, would make an artificial selection based upon the evaluation of the products, which, when performed iteratively, would correspond to serial transfer. In this way we are no longer dependent upon natural selection, tied as it is to the efficiency of reproduction, but can apply our own, chosen criteria. The method is equivalent to a reconstruction of the value landscape (Vignettes 10 and 12). The computer designs, as it were, a contour map of the value distribution of the cloned molecular ensemble. To make a contour map, one must know the exact coordinates and height of each point. Geographical coordinates of position can be transformed into coordinates of distance. These correspond to the mutation distances between the clones, based on a hierarchical scheme. The heights are in our case represented by fluorimetrically registered function values. They can be tuned to specific properties of the phenotypes produced in the clones by the chosen test procedure.

The apparatus shown in this vignette was developed at the Max Planck Institute for Biophysical Chemistry in Göttingen.[60] An evolution machine of the kind just described is currently under development at the same institute.[60] It incorporates evolution into traditional biotechnology.

12. Sequence space[20,38]

Our task here is to find a way of illustrating spatially the sequences of nucleic acids and the relationships between them. Closely related sequences must be given coordinates close to one another, and it is not enough for the representation to encompass the mutants close to one particular sequence, such as the wild type; it must be able to represent distances between *all* mutants taken pairwise. This cannot be achieved with planar solutions, in which, for example, the wild-type sequence stands in the middle and the mutants with k errors are placed around it in concentric circles of radius k. The idea of sequence space can only be realised if the 'space' used is multidimensional. The following illustration is intended to show why this is so. We consider a binary sequence of symbols and build up a sequence space for it.

We start by considering just one position in the sequence, and this can be occupied in two ways. The sequence space therefore consists of two points, one representing occupation by R and one representing occupation by Y. We connect these two neighbouring points by a straight line, which is the one-dimensional sequence space.

If we now add a second position, which also can be occupied in two ways, the effect is simply to duplicate the first line. Connecting the ends of the two lines that correspond to one another, we obtain a square, the two-dimensional sequence space. If there are three positions, then duplicating the square leads to a cube (three dimensions) and so forth. It becomes clear how the sequence space is built up: each new position calls for the duplication of the previous diagram, and the new sequence space is formed by joining the points of the preceding space to the corresponding points in the duplicate. In this way, all points that can be reached by one mutation are connected by a line. For a binary sequence containing ν symbols, the result is a hypercube in n dimensions, with each point having n nearest neighbours. The idea of sequence space was developed in information theory by Richard W. Hamming,[61] and the possibility of its application to nucleic acids and proteins was first suggested by Ingo Rechenberg[21] and John Maynard Smith.[21] The methods for comparing sequences described in Vignettes 1 and 2 are based upon the idea of sequence space.

Although we cannot readily picture a multidimensional structure, we can imagine some of the properties of the six-dimensional hypercube in our chosen example (see illustration).

- The first thing we notice is the enormous capacity of the hypercube, which grows exponentially with the number of positions in the sequence. Each new position doubles the volume. To illustrate this, let us consider our universe,

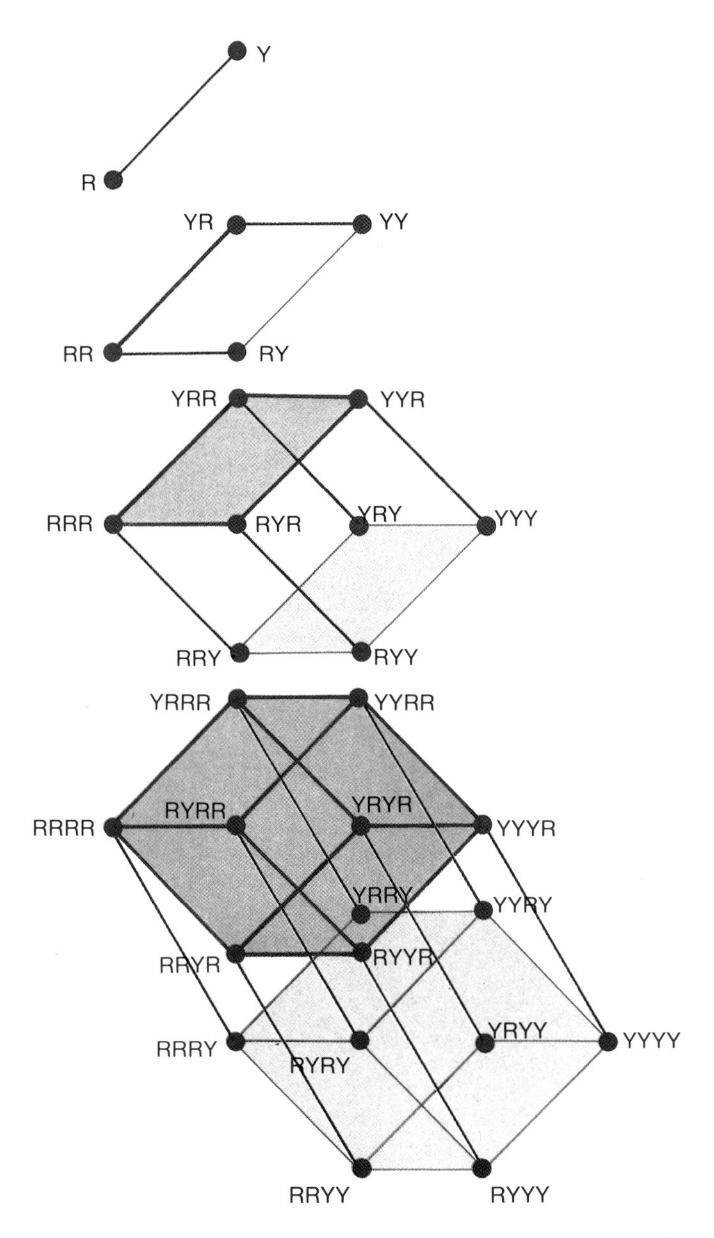

which has room to accommodate around 10^{108} hydrogen atoms.* If we assigned one hydrogen atom to each point in sequence space, we could place all these hydrogen atoms within the hypercube that corresponds to a binary

* For this calculation, we have taken the universe to be a sphere with a radius of ten thousand million light-years, and the hydrogen atom to have a volume of one cubic Ångström unit.

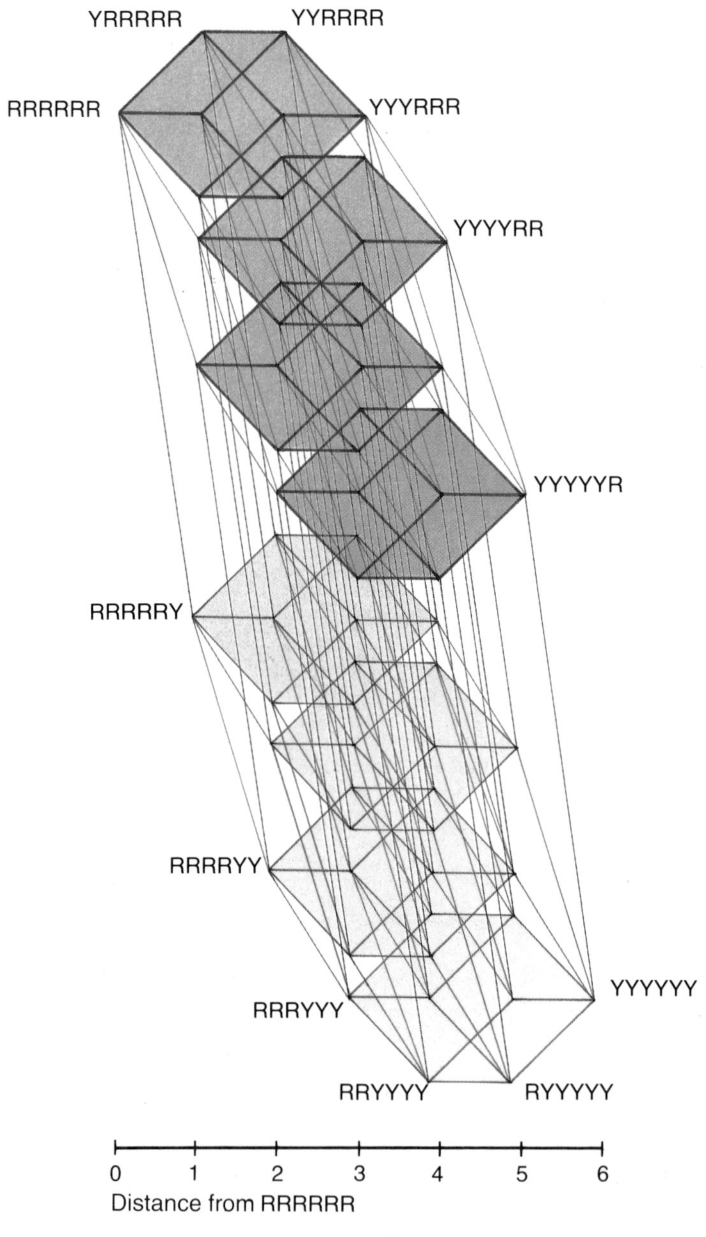

sequence consisting of 360 symbols ($2^{360} \approx 10^{108}$). In other words, the volume of our universe expressed in Ångström units corresponds to the number of vertices of the hypercube associated with a 360-membered binary sequence. Yet most of the nucleic-acid sequences with which biochemists deal are much longer, and are not merely binary.

- In spite of the gigantic volume of our hypercube, the detour-free distance between two points never exceeds the dimension ν (in our example above, 360). Since sequence space has no scale, the distances do not have units. If one asks the way in New York, the answer will usually include a distance given in street blocks, rather than in miles. In the same way, we define the mutation distance as a pure number, the number of the positions that are occupied differently in two comparable sequences. The fact that distances are small in sequence space leads to an important consequence. Any goal can be reached quickly, as long as a 'street map', or some other sort of guidance, is available. Without this, however, one would get just as hopelessly lost as in the far-flung galaxies of outer space.
- The number of connecting lines between the points is, as the diagram makes clear, confusingly large, in spite of the fact that the points are connected only to their nearest neighbours. Each line is a single mutational jump; if we also drew lines corresponding to mutation at two or more positions at once in a single copying process, then the number of lines connecting any given point with other points would soon exceed the population number itself.

All these aspects will appear in even sharper relief for the sequence space associated with a sequence consisting not of two but, like the nucleic acids, of four symbol classes. How can such a space be imagined? We can attempt this by starting with a binary sequence space and representing the positions in the sequence by their base classes R and Y. Each point in this space now corresponds to a particular sequence in RY notation. Next, we give the bases their correct identities. This means a further binary decision for every position, as there are two sub-categories for R (A and G) and two for Y (U and C). In other words, every point in ν-dimensional binary sequence space is allotted a ν-dimensional subspace. The overall dimension is thus now 2ν, and we have a total of $2^{2\nu}$, or 4^{ν}, possible states. The figure shows a sequence space of this kind for $\nu = 3$. It shows an alternative perspective on the scheme of the genetic code shown in Vignette 8. The eight binary sub-spaces of the binary RY space are coloured appropriately.

The evolution of forms of life corresponds to small, stepwise changes and sometimes to large jumps in the sequences of genes. Since each sequence is represented by a point in sequence space, and since the relative positions of all points reflect the neighbourhood relationships between the sequences, any evolution process must be describable by movement along a connected route through sequence space — a movement analogous to diffusion. Selection can be regarded as a 'condensation' in sequence space (Vignette 10). The sequences represented in the population are not distributed randomly throughout sequence space, like molecules of a gas; they are located in a relatively narrowly defined region. In the evolution of a gene, the centre of mass of the localized distribution diffuses around in the sub-space allocated to the particular gene. The route is influenced decisively by the contours of the value landscape.

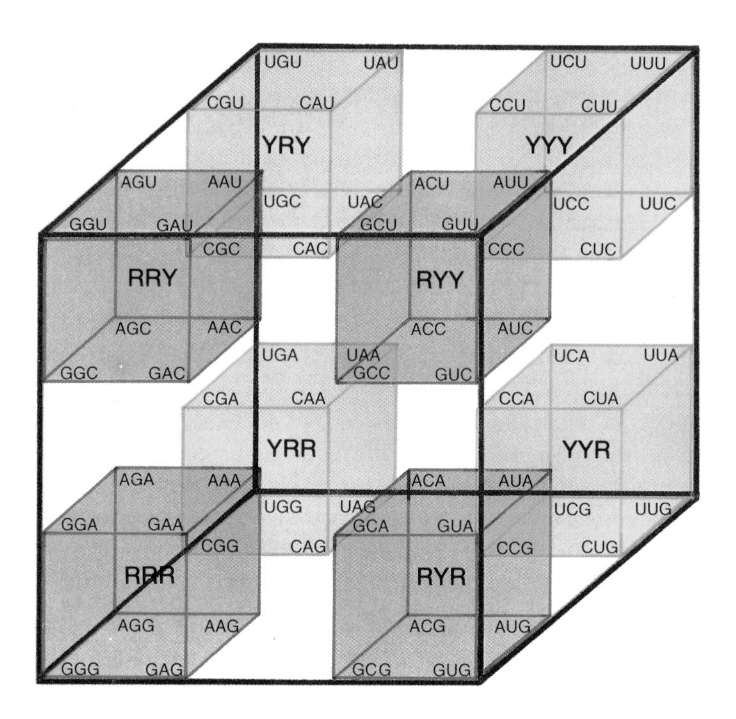

Condensation always occurs at the top of the highest local peak. The mutant distribution occupies this and other neighbouring value peaks, and tries to move along the complex ridges of ν-dimensional space towards the highest peak.

In the classical picture of Darwinian evolution, a gene continues to mutate until, by chance, a better-adapted variant arises. This becomes selected, and the procedure continues, so to speak, one level higher. The enormous number of possible states means that random moves of this kind can only lead towards a goal if there is a monotonically (i.e., uniformly) rising path in the value landscape that, projected on to one dimension, looks something like this:

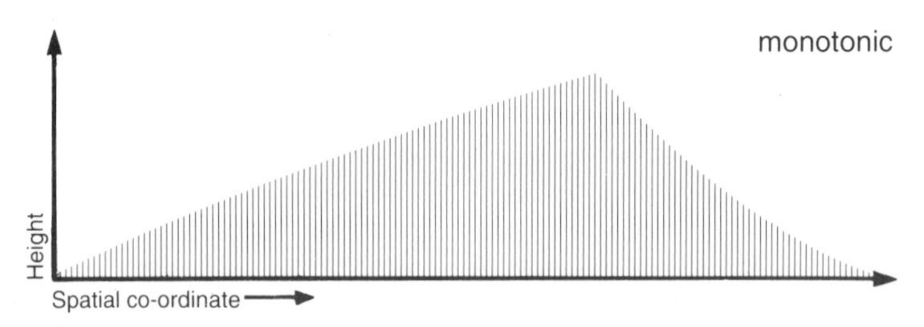

We know that this idea is not realistic. However, the other extreme of a completely structureless value landscape is equally unreal; the heights are not arbitrarily distributed. To see this more clearly, contrast the profile of a feature on the surface of the Earth with a random distribution of altitudes:

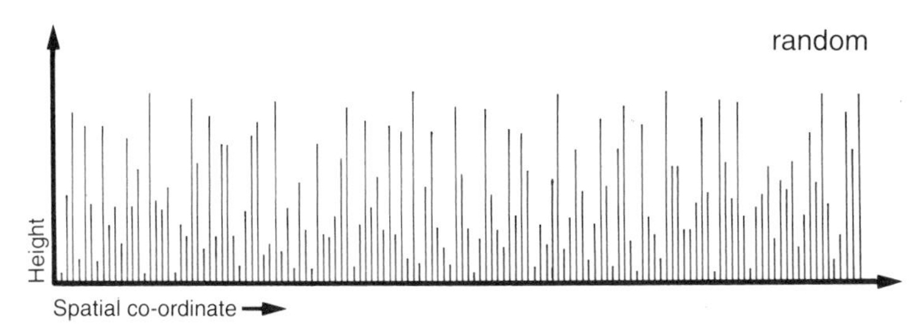

Benoit Mandelbrot[62] has shown that the distribution of altitudes on the Earth's surface can be represented by *fractal* structures. Fractals have the same appearance, irrespective of their scale; they are therefore termed 'self-similar'. Mandelbrot called a landscape structure 'Brownian' if it can be generated by a Brownian diffusion process. A Brownian process can be simulated by tossing a coin[8]. If in a long series of tosses the aggregate number of 'heads' minus 'tails' is plotted against the number of throws, then the result is a Brownian landscape as defined by Mandelbrot:

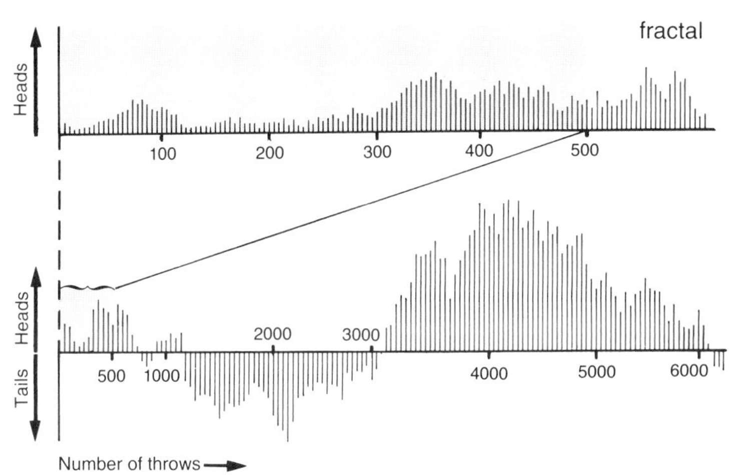

The self-similarity of this landscape becomes evident when we compare the upper plot with the lower. (The lower plot represents, on a reduced scale, ten times as many throws as in the upper plot.)

Mandelbrot considered a further model, which we might well regard as the converse of the evolution process. He asked: What chance has a raindrop that lands on a hilly island of reaching the sea? The profile of the hills is shown below. If the island is one-dimensional, as the diagram suggests, then the raindrop has no option but to wait until all the intervening hollows have been filled up and then to flow out over the resulting 'terraced landscape':
Analogously, an upward evolution process can only proceed along a ridge if this is free from local peaks that necessitate a brief descent along the way.

The solution to this dilemma lies simply in the fact that landscapes on Earth

are not one-dimensional. The existence of two dimensions (apart from altitude) turns depressions into hollows that are spread out over the two dimensions, the edges of which also have a distribution of heights. If we take an arbitrary vertical cross-section, there is apparently a barrier to be overcome; but viewed in two dimensions this barrier is a mere point on the edge of the hollow, and other such points may be higher or lower. Only one of these points can be the lowest (see cross-section 3 in the illustration below). For the others, there is always a lower point at the edge of the hollow, from which water can flow out. So the hollow does not need to be filled up to the height of the point in whatever cross-section we take, but just to the lowest point of the hollow's perimeter. In other words: raising the number of dimensions increases the number of possible routes, and, if the dimensionality is high enough, the raindrop will always find a way to the sea.

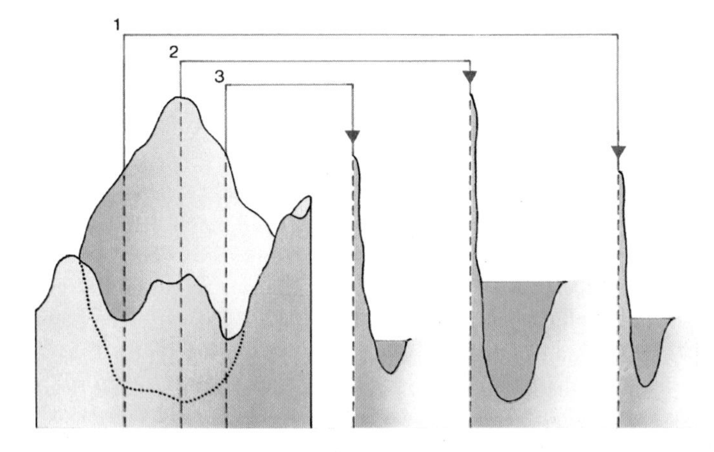

The situation is similar for the reverse process, the evolutionary ridge walk. However, we should bear in mind that sequence space is 'coarse-grained'; it is a space of points. So the altitudes vary in small jumps, not continuously.

Furthermore, mutation jumps are possible, so that from each point 10^{10} or more points within reaching-distance can be scanned. So evolution in multi-dimensional space has a very good chance of finding a continuous route of small steps to the highest peak of the value landscape. The value contours are reflected in the population topography (see Vignette 10).

However we try to visualize the fitness landscape, we must always remember that we at no point are further than ν steps away from the highest peak (where ν, as before, is the sequence length). As there are multiple connecting lines — corresponding to the many possible mutation jumps — the chance of finding the ridge leading to the highest peak is much greater than it would be if we just wandered around at random. In Vignette 10 we have given an example to show how this aid to orientation can accelerate the process of evolution by many orders of magnitude. Perhaps the following picture is the most apposite way of describing the fitness landscape: if the multidimensional hypercube were transparent, then we would always be able to see mountains in our immediate vicinity, even though there is always a vast number of points and dimensions at zero altitude separating us from them.

13. Viral infection[31,32]

Nature's closest counterparts to the replicator are the viruses. These stand right at the border between the living and the non-living, and this makes them especially suitable as models for the earlier stages of evolution. They are in all probability late arrivals, having arisen out of autonomous species; many believe that the viruses originally were parts of the functional programmes of their host organisms that escaped, established themselves, and developed independently. This would at least explain the intimate inside knowledge of their hosts that the viruses appear to possess. Their genetic programmes code only for a few functions, with the help of which the mechanisms of the host are redirected towards the reproduction of the virus. As replicators, the viruses are subject to natural selection. The limited length of their genomes and their correspondingly high rate of mutation enable viruses to adapt very rapidly to a new environment. We have seen in Vignette 2 that the influenza virus evolves many millions of times faster than do autonomous microorganisms.

From the many different species of virus, we shall examine in more detail the single-stranded RNA viruses, which are also models of pre-cellular RNA systems. We distinguish three types, on the basis of their mechanisms of infection: plus-strand viruses, minus-strand viruses, and retroviruses.

Plus-strand viruses

Their name indicates that the genome of this virus itself carries information; it is accepted directly by the host cell and translated directly into protein. The mere introduction of this RNA into the cell is enough to initiate infection.

We can take as an example the bacteriophage Q_β, whose host is the gut bacterium *E. coli*. The quantity of information that it can carry is restricted by its relatively high error rate in replication (3×10^{-4}), and this information must be used economically. With viruses of this kind, economy is often achieved by making the genes overlap. The Q_β genome encodes proteins with four different functions:

- a coat protein, which protects the viral RNA from hydrolysis;
- an adsorption protein, for adhesion to the host cell;
- a replication factor, which promotes virus-specific RNA synthesis and thus the exclusive replication of the viral RNA;
- a lysis factor used to disrupt the host cell after multiplication of the virus.

The most remarkable part of the life cycle of Q_β, and the most important for its evolutionary success, is its highly specific mechanism of multiplication. For this purpose, the virus does not even provide a complete enzyme. It is well acquainted with the contents of the host cell, and it simply employs three proteins of *E. coli* that normally have a completely different function connected with protein synthesis. These three proteins become yoked to the virus-specific replication protein and form a complex with four sub-units that replicates viral RNA with a high activity and specificity. For the RNA of the host cell, this complex is completely inactive. Only the RNA of the virus possesses the password that opens the way for it to be copied, and this copying takes place in so uncontrolled a manner that the *E. coli* cell ultimately dies of it. There are indications that the virus-specific factor has a precursor in *E. coli* that changed its function during the evolution of the virus.

Looked at in detail, the infection cycle proceeds as follows.

- The host cell is recognized by the virus, which possesses an adsorption protein especially for this purpose. This protein binds to the host cell's F-pili, which are whip-like structures on the cell surface; the virus then injects its RNA into the cell.
- The viral RNA, once inside the cell, is accepted as mRNA and translated. The products of translation include the replication factor, which binds to the three cellular proteins (above) so as to form the replication enzyme.
- With the help of the replication enzyme, the viral RNA now multiplies many times over, while the host's RNA is unaffected. The original single copy of viral RNA becomes the progenitor of between 10 000 and 100 000 copies. Most of these are then packed into the coat proteins that also have been synthesized in the meantime.
- The lysis factor then causes the bacterial cell to burst. About 10 000 virus particles are released. However, only a certain fraction of these are infectious; this is because the high error rate leads to a broad distribution of mutants (compare the quasi-species), and there is no selection pressure acting upon the viruses within the *E. coli* cell.

Further examples of plus-strand viruses are the majority of the plant viruses, such as the tobacco mosaic virus. They also include several pathogenic viruses of mammals, among these the picorna viruses, a class that embraces the viruses causing poliomyelitis, hepatitis A, and foot-and-mouth disease.

Minus-strand viruses

This group has a genome consisting of a single-stranded RNA molecule complementary to the information-carrying strand that the host cell will translate. These viruses must therefore bring their own replication enzyme with them into the host cell, in order to make the first complementary copy upon which the first viral proteins are synthesized.

We shall take a look at the mildly pathogenic virus that causes the disease vesicular stomatitis in cattle. Its genome contains around 11 000 nucleotides and codes for five proteins, two of which are subunits of the virus's replication enzyme. The cycle of infection runs as follows.

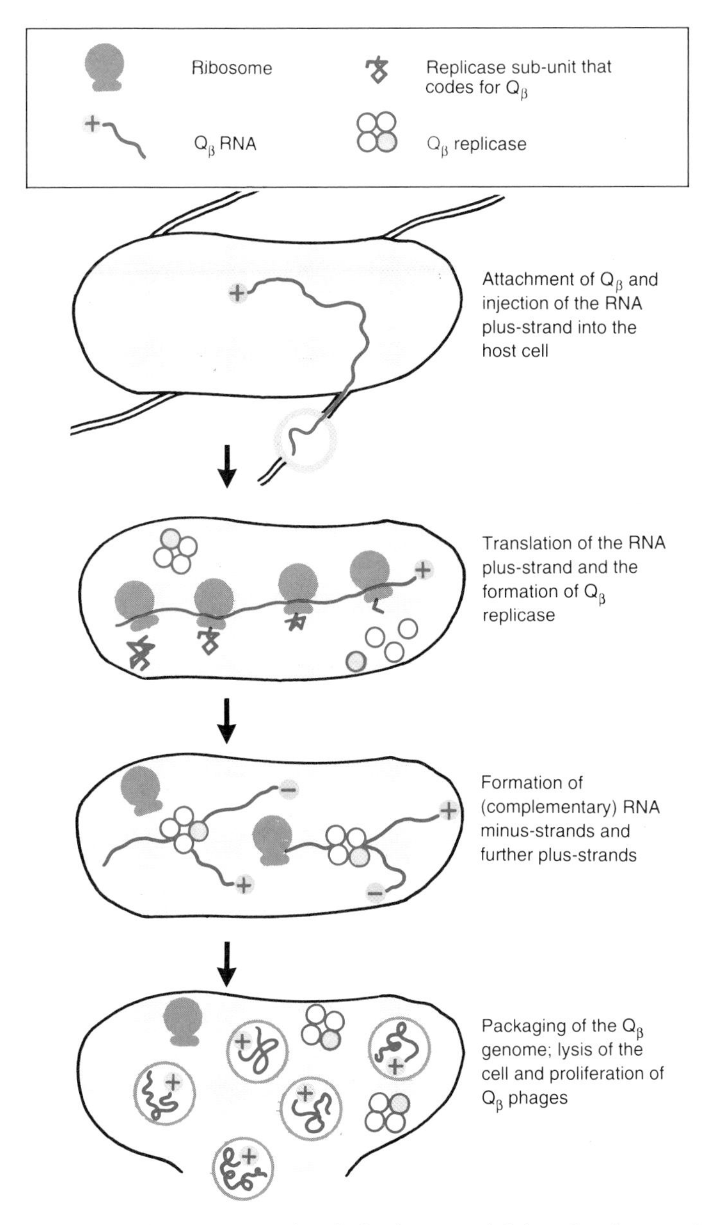

First, the virion (i.e., the complete infectious particle) makes its way into the host cell. The replication enzyme it has brought along with it transcribes the minus strand into five separate mRNA molecules, and these are translated into protein by the host cell. In this way, the functional units of the virus come into being. The rest of the infection cycle resembles that of the plus-strand virus. However, an important feature of this is that, in addition to the five mRNA molecules, entire viral genome strands are copied, both from minus to plus and from plus to minus. These are needed for the proliferation of the virus, in which a package is made consisting of an RNA minus strand and one copy each of all the five proteins. These are: two replication proteins, the coat protein, an antigenic surface glycoprotein and a membrane protein. The completed virus is packed into a lipid membrane, the lipid components of which are also taken from the infected cell.

The virus described belongs to the class of the rhabdoviruses. Other minus-strand viruses include the rabies virus and the influenza virus, whose gene consists of eight single RNA strands with a total of 14 000 nucleotides and which encodes three replicative proteins.

Retroviruses[51]

The name of these viruses bespeaks an essential aspect of their life cycle: the information in their RNA is copied back into the DNA of the host. This is done by a special enzyme called *reverse transcriptase*. The reverse transcriptase turns the single strand of viral RNA into a hybrid containing one strand of RNA and one of DNA; the DNA then separates off and is copied again by the reverse transcription enzymes, giving a double strand. This is then incorporated into the genome of the host cell by a multistep procedure, which can be illustrated as follows.

At both ends of the 'viral DNA', there are identical sequence segments with a signalling function. Not only do these contain the 'start' and 'stop' instructions for replication, but they also control a splicing and cutting function, and when the synthesis of the double strand is complete the viral DNA becomes circularized.

The circular viral DNA now seeks out a particular structure in the host DNA which it can recognize. The DNA strands of the virus and the host are cut at this point, and the ends of the viral DNA are spliced into the gap formed in the host DNA. The viral DNA spliced into the host is termed a *provirus*. In general, the host cell is not destroyed by the presence of the provirus, and the provirus multiplies along with the cell in which it resides. The virus may, however, transform the host cell, and it may in fact cause tumour formation.

Another retrovirus is the highly pathogenic virus HIV, which causes the immune disease known as AIDS. Its target is the command headquarters of the immune system, which, as a consequence of the infection, is put out of action some time later. This virus has a high degree of immunological flexibility, as a result of its quasi-species nature (see Vignettes 9 and 10).

In addition to the single-strand RNA viruses described here, there are also double-strand RNA viruses and both single- and double-strand DNA viruses. An interesting class of RNA viruses is the reoviruses: these contain ten individual double-stranded RNA molecules that code for ten different proteins. The eucaryotic host cells of these viruses are unable to decode mRNA sequences that code for more than one protein molecule, so the reoviruses have adapted to this constraint.

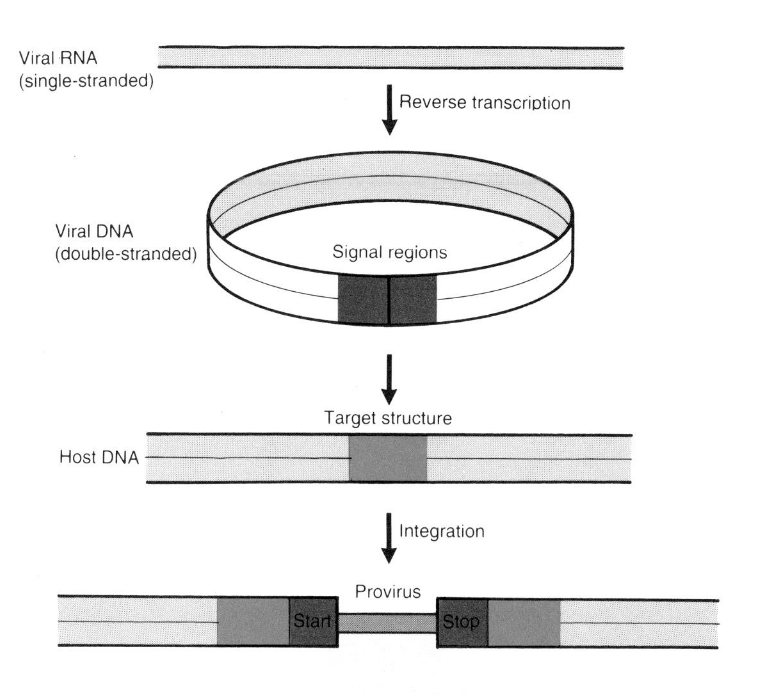
Viral RNA
(single-stranded)
Reverse transcription
Viral DNA
(double-stranded)
Signal regions
Target structure
Host DNA
Integration
Provirus
Start
Stop

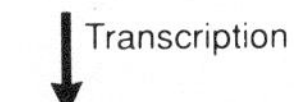
Transcription

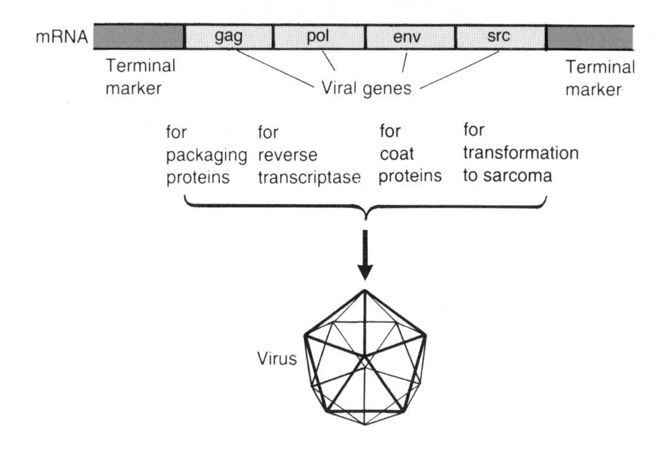
mRNA
gag
pol
env
src
Terminal
marker
Viral genes
Terminal
marker
for
packaging
proteins
for
reverse
transcriptase
for
coat
proteins
for
transformation
to sarcoma
Virus

Single-stranded DNA viruses can be compared with the plus-strand RNA viruses. The phage ϕ_x contains a DNA genome of about 6000 nucleotides. Double-stranded DNA viruses vary considerably in size. The polyoma virus (that causes tumours) contains only five genes; the T_4 phages, in contrast, have over a hundred, and are thus only about an order of magnitude smaller than their host, the bacterium *E. coli*[63]. Their error rate is adapted to the size of their genomes, so that while they are much less variable than the RNA viruses they still evolve much more rapidly than the autonomous microorganisms can. The T phages also contain genes for tRNA. Their sequences reveal a very close kinship with their congeners in *E. coli*, but they also show a greater amount of sequence blurring. This demonstrates once again the close relationship between phage and host cell.

14. Hypercycles[12] and compartments[34]

Plus- and minus-strand viruses have revealed two principles of organization that we encounter again and again in Nature: cyclic reaction pathways and compartment formation. It is likely that these principles once levelled the way from gene to genome and from genotype to phenotype.

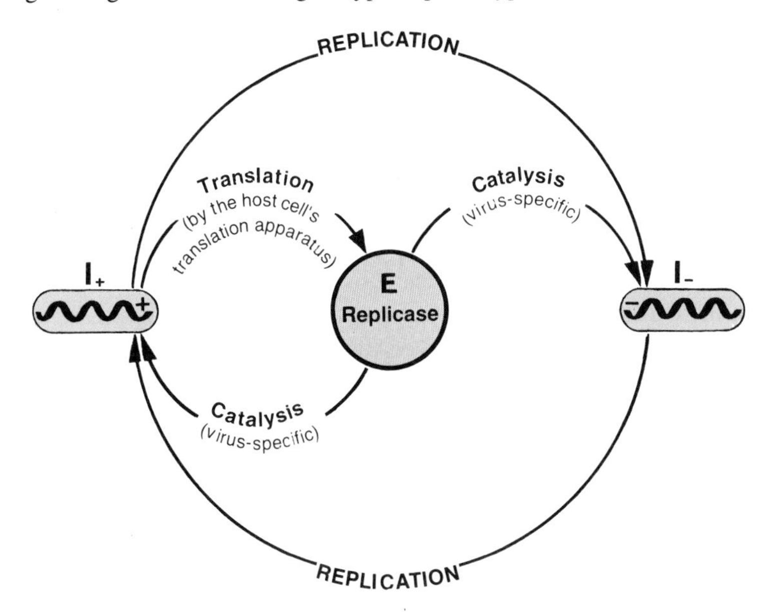

Let us recall the replication cycle of the plus-strand virus (Vignette 13): genotype and phenotype are no longer structurally identical. The information resides in the RNA strand, in the genotype. Only the genotype has the ability to set off the chain of events inside the host cell that lead to its own reproduction and (thereby) to its preservation. Selective evaluation is carried out both on the viral RNA, with its phenotypic characteristics, and on its translation products, including in particular the protein subunit responsible for the specificity and catalytic efficiency of the replication enzyme. This genotype-phenotype dichotomy is spanned by cyclical feedback coupling. A mutation in the genotype that expresses itself in the phenotype brings about an immediate evolutionary response. If the change is favourable for the replication enzyme, then the new genotype will reproduce more rapidly and be selected in preference to its predecessor. If the mutation is disadvantageous, the precursor will remain dominant. We call a reaction cycle with superimposed higher-order cyclical coupling a *hypercycle*. What does this term imply?

A simple catalytic cycle might be as follows: plus-strand is the template for minus-strand, and minus-strand is the template for plus-strand. If the replication of such a cyclical system is catalysed by an enzyme already present in the surrounding environment, then its essential nature is unaffected. If, however, the replication enzyme is encoded in one of the two strands (by definition, the plus-strand), then the system is no longer a mere catalytic cycle, but a hypercycle. It would make no difference if one of the RNA strands was itself the enzyme (a 'ribozyme'). In either case, the overall rate of the reaction would depend upon the concentrations of both template and catalyst. Since, in general, a reaction rate is directly proportional to each of these concentrations, the rate is in this case determined by their product (template concentration times catalyst concentration). Therefore, irrespective of whether the catalyst is one of the replicators or one of the translation products, the rate of the reaction (in the absence of other regulating mechanisms) will rise with the *square* of the RNA concentration. This means that the increase in RNA concentration as time passes is not exponential, as with simple replicators, but instead is hyperbolic. The case just described is the simplest hypercycle. The amplification of viral RNA and protein inside the host cell is indeed governed by a hypercyclic mechanism, as has recently been demonstrated for the bacteriophage Q_β[32]. In Nature, there exists an entire class of more complex reaction networks that we can term hypercycles, and these may utilize both positive and negative feedback control. An example of positive feedback is enzymic reinforcement, and negative feedback can be provided by repressor molecules. The hypercycle represents a form of organization that, like the quasi-species, displays qualitatively new properties.

The feedback loop that connects the replication enzyme to its RNA template can only come into effect if the genotype and the phenotype remain in each other's vicinity — in other words, if they are encapsulated in some kind of compartment. Only then can the phenotype act upon its own genotype and not upon genotypes of other, competing replicators. This principle of compartment formation, or *compartition* of the reaction medium containing the quasi-species, is the second principle spanning the phenotype–genotype dichotomy. It can clearly be seen at work in the case of the minus-strand virus. Its genome is not infectious in itself. It has to bring along its own reproduction enzyme, and it can only operate when both genome and enzyme, components of the virion, are present in the same compartment. We see the same principle at work, even more clearly, in the influenza virus, with its genome consisting of eight separate RNA molecules.

Can either of the two organizational principles, hypercycles and compartments, be dispensed with, or replaced by the other?

Let us return for a moment to the beginning of evolution. Evolution means optimization, and is associated with selection. Selection, in turn, is a direct consequence of replication. Evolution began with the appearance of the first replicators. However, the error-threshold relationship severely restricted their length, on account of the relatively low accuracy of their replication. The path from the single gene to the genome complex demanded the integration of replicators. Yet, if we consider the complexity of the genetic code and its associated machinery of translation, we can clearly see that even for a primitive code a whole arsenal of functions is needed, all of which must first be encoded in the genes that are to be integrated. In order to satisfy the error threshold relationship, the accuracy of replication must be sufficiently high. It should be at least as high as in present-day viruses (with their mere four to eight functions). This in turn must have necessitated highly developed copying enzymes. In other words, what should have been the goal of the evolution process was already indispensable at the start!

Thus, because of the error threshold relationship, the most primitive genes that were translated could never have begun as an integrated genome. At first, they were forced to cooperate as separate entities. The integration of these isolated genes could be brought about by hypercyclic coupling or by enclosure into compartments.

Our thesis will be that both forms of organization are needed at once, and to show this we shall now compare their properties. We consider first the hypercycle, which is the cyclic coupling of individual replication cycles, represented here in a generalized form.

The vital point here is that the coupling is *cyclically closed*, so that the feedback loop embraces all the individual members. We shall not attempt here to discuss the detailed mechanism of such a coupling, which could raise questions such as whether reproduction is actively catalysed (promotor effect) or merely de-repressed (reversal of inhibitor effect), or the issue of whether catalysis is carried out by an RNA strand or by the strand's translation products, or indeed by environmental factors influenced by the products of translation.

The hypercycle is one of many reaction models whose principles can be derived from naturally occurring systems. It is a category of reaction networks with deducible characteristics. Thus, we are not looking for historical details of the origin of life, but rather raising a question of physical chemistry, when we ask: What can hypercycles do, and what are their limitations?

The hypercycle fulfils the following requirements.

- The individual participating replication cycles — genes or polycistronic gene sequences — operate close below their error threshold; they are thus protected from decay by error accumulation, and they satisfy the selection criteria for quasi-species.
- The competition between different replicators is put out of action by the cyclic coupling, and replaced by mutual cooperation. The concentrations of the replicators participating in the cycle reach steady, stable values. If the concentration of a particular catalyst rises, then the reaction it catalyses is accelerated, and the result is a raising of the concentrations of the other replicators that follow it in the cycle. Any advantage that one partner gains,

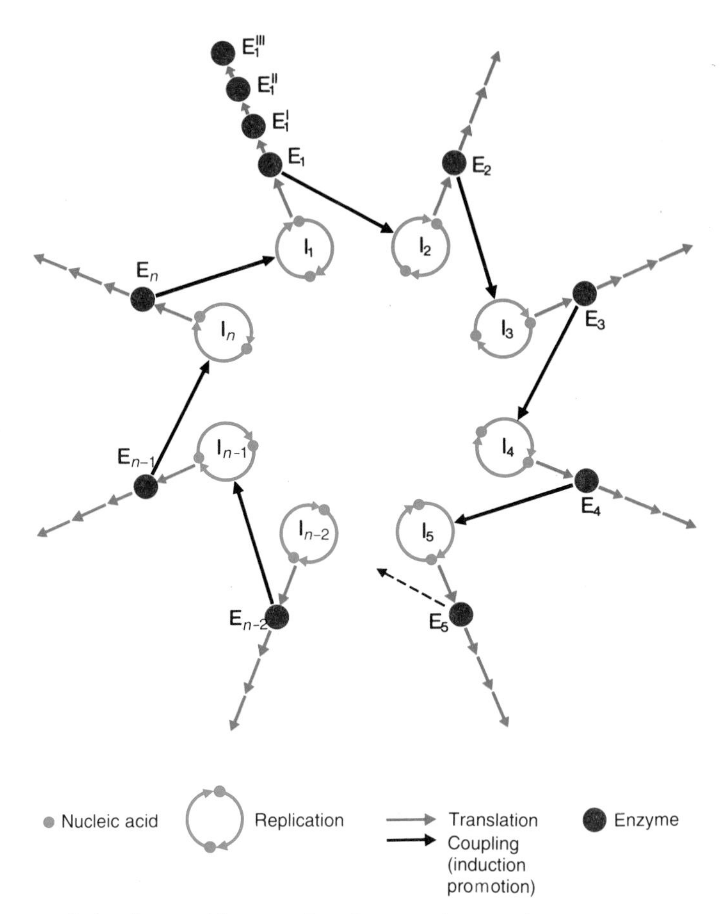

for example by favourable mutation, is passed on to the next partner, and so on. So every advantage is shared by all the members of the cycle, and these grow up in a synchronously regulated way.

The entire hypercycle participates, as a single unit, in the selection competition. Here the selection is much sharper than in the case of simple template-instructed replication, because of the quadratic rate law governing the growth of hypercycles. The result is a 'once and for all' decision; a system that has been selected can be improved, expanded, or rationalized, but it cannot easily be displaced by an alternative, competing system. This behaviour can no longer be called selection in the Darwinian sense. However, such 'once-and-for-all' decisions have indeed been made during the early period of evolution. Both the genetic code and the biosynthetic mechanism of the cell are universal, with only very minor divergences that are explicable. Chirality, too, is uniform among biological macromolecules. All these decisions were taken during the development of the translation apparatus, at a point in time when the emergence of the genotype–phenotype dichotomy made the appearance of hypercyclic feedback a matter of inevitability.

The hypercycle has three properties of crucial importance. First of all, it unites several genes that are working just below their error limit, and thus bypasses the error threshold, allowing the quantity of information to rise to the much higher levels needed for the nucleation of an apparatus of translation. Secondly, it controls cooperation by adjusting the concentrations of each reaction partner to suit the others. Thirdly, it tolerates no competition, spreads rapidly, and can evolve as a unit by retaining favourable mutations and expanding or contracting in response to selection pressure. However, this facility is not unrestricted, which brings us to the weaknesses of the hypercycle.

The hypercycle can only retain those mutations that improve the mutant in respect of its function as a member of the coupled system. Mutations that alter the active function of a member cannot be selected for or against. (Active function could for instance be catalytic activity of any of the phenotypes, while passive funtion is coupling, such as binding, between genotypes and phenotypes.) For selection of this kind, the mutant would have to compete with its own precursor and it cannot do this, because in the hypercycle it is not itself the target of selection. This can have catastrophic consequences for the hypercycle, if the passive function is optimized at the expense of the active function: whenever this occurs, the hypercycle ultimately destroys itself. A hypercycle can perform coupling, functional integration and growth regulation, but it cannot maximize the functional efficiency of its members and is therefore perpetually in danger of degenerating.

Let us now look at the second of the organizational principles that makes integration possible: compartition. An excellent example of this is provided by the influenza virus. With its minus-strand RNA genome, it is obliged to carry its own replication enzyme around with it. Whenever a new, functional virion is produced, all the partners must be correctly packed in. From this fact, one can derive the following compartment model.

Compartments offer the following advantages.

- Assuming that the mechanism guaranteeing correct composition functions in the same way as in the case of the influenza virus, so that only a fraction of the daughter compartments contains the correct assortment of molecules, active functional advantages can be selected for. The compartment whose contents corresponds to the best mixture of mutants of all eight RNA components will have the highest rate of proliferation and will be selected.
- Enclosure in a compartment reduces the effects of the dilution of valuable mutants. It is not necessary for the entire medium to attain a particular concentration of the appropriate components; concentration conditions allowing efficient cooperation can be attained inside the compartments.
- The concept of a compartment does not need to be formulated as narrowly as

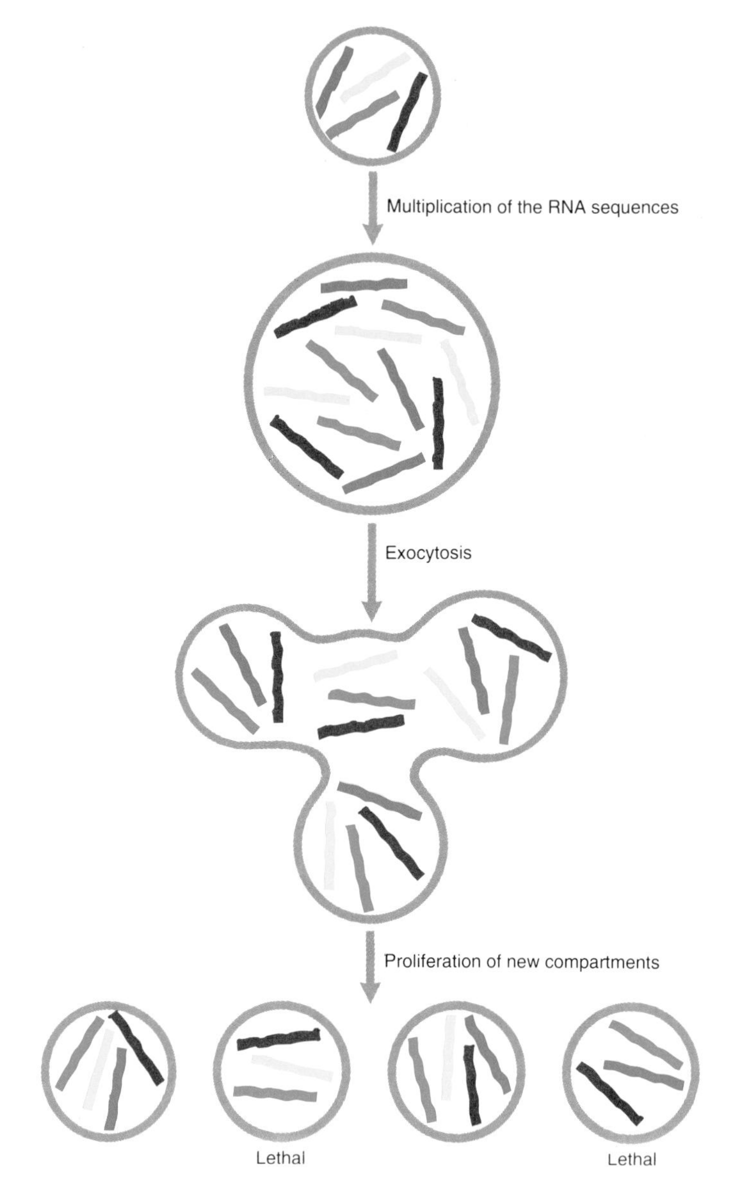

it was in our discussion of viruses. A compartment can be anything from a lipid vesicle with the capacity for endo- and exocytosis, or with an open surface, to a DNA/RNA strand with an exon/intron structure, or a complex, or an enrichment of concentration at a surface, or even a mere neighbourhood interaction with restricted diffusion away.

However, these advantages are qualified by a number of defects sufficiently serious to prevent the organizational principle of compartition from being sufficient to bring about the integration of genes.

The most conspicuous disadvantage is the complete absence of any organized cooperation among the replicators. In fact, in the confined space of the compartment, there is even stronger competition among the replicators than in the free solution. The strong selection pressure that results will, just as in our evolution experiments, lead inevitably to an 'either–or' selection. The component that grows the fastest overruns the compartment and prevents a balanced distribution of components from being passed on to compartments of the next generation. In a compartment containing competing species, the law of the jungle applies: the stronger a component is, the more it profits. In contrast, the components of a hypercycle in a compartment automatically adopt a balanced distribution, since the weaker partners profit from the stronger ones, which give more than they take.

Calculations[64] have shown that evolution can still take place in a compartment under special conditions. We can, for example, assume that only those compartments with a complete set of components can survive, and that all mutants are either well adapted or lethal. The surviving compartments then have an infinite selective advantage over the others, so that the error threshold relationship loses its meaning. If, however, there can occur mutants with all possible degrees of fitness, then we expect to find arbitrary mutant mixtures with almost continuous fitness spectra. The error threshold relationship will then impose the same restrictions as in the case of the quasi-species, and the accuracy of reproduction must be adapted to the total amount of information in the compartment. But if this were the case, then the integration of different replicators by compartments would have no advantage, since the same effect could be obtained simply by integrating all the genes into a single extended genome. Compartition would have made no contribution to overcoming the error threshold.

The comparison of hypercycle and compartment shows that the two principles of organization complement each other ideally. The one is weak where the other is strong. What could be more logical than to combine them in a 'hypercyclically organized compartment'? The replicators in this combined organizational form could be given very different functions and still, originally, have been mutants of one and the same quasi-species. Once a coupling between genotype and phenotype has arisen, it is more or less normative for all mutants. This can only lead to a highly branched reaction network and the systematic formation of feedback loops. This is of value for all the components of the loop. In such a system, the hypercycle can multiply and come to dominate the scene. In a compartment, the hypercyclically organized replicators become adapted to their new tasks by mutation and selection. The route to hypercyclicity is predetermined for the quasi-species in the compartment: the hypercycle does not need to arise *de novo*, and it can evolve. This leads to a considerable expansion of the information content of the quasi-species, to the establishment of new mechanisms with the help of which the error threshold can be raised, until ultimately structural integration can take place, leading to a DNA genome.

In the later stages of their evolution, the viruses have traced out the same path. They, too, are more than mere compartments. The infection cycle of the influenza virus is a subtly contrived process, in which the eight RNA genes come into action at the right time and in the correct order. The individual phases of this process are only just beginning to be understood. Nature has an abundance of material for study that is waiting for us with many further surprises. The more we can learn about the molecular mechanisms of present-day living systems, the better we shall come to understand the processes of early evolution.

15. Recombinant DNA[65]

The genes of almost all organisms are today integrated into huge DNA molecules. From the one-time compartments with organized gene ensembles have arisen cells with centralized legislative and regulated executive, fundamental units of life, molecular communities. Although they are penned in from their environment by well-defined boundaries, they are in perpetual contact with it, making use of a highly sophisticated system of transport and communication. The integration of genes to give a genome has facilitated the synchronization of gene duplication and the division of their compartment. It has turned the cell into a replicator unit able to diversify its structure and to specialize its role. Mechanisms of error correction allow the information of a cell to be transmitted *en bloc*. The age of 'once-and-for-all' decisions, which led to the establishment of a universal code and to general biochemical mechanisms, is long gone.

Yet, as efficient as the molecular machinery of the first microorganisms became, its size made them correspondingly inflexible in their ability to evolve further. One reason for this was the high accuracy of genetic replication that they had attained; another was the fact that changes could only be passed 'vertically' down lines of descent. This may indeed be a reason why evolution tarried on the level of single-celled organisms for more than two thousand million years, even though these had taken less than half this time to develop fully. The subsequent phase of evolution, that spawned the millions of different species alive today, has taken only a few hundred to a thousand million years. The way out of the unicellular cul-de-sac required a new principle of organization, a principle that made 'horizontal' gene exchange possible: genetic recombination and transposition. It dates back to very early in the unicellular phase of life.

Homologous genetic recombination (see the following figure) is the basis of sexual inheritance. Any exchange of genes requires a set of tools for cutting and splicing DNA, and a system of book-keeping for the DNA involved. The product of genetic recombination must be a complete set of genes and not just a random mixture. We encounter recombination at almost all levels of life. It is an invention of DNA, one that must have been inevitable as soon as genetic information had become embedded in giant molecules containing many genes. The transposition of genes makes possible the creation of new genomes, and it must therefore always have had central importance in the evolution of organisms.

The scheme of genetic recombination proceeds via the intermediate stage of a *duplex double helix*; this has been observed directly in the electron microscope.

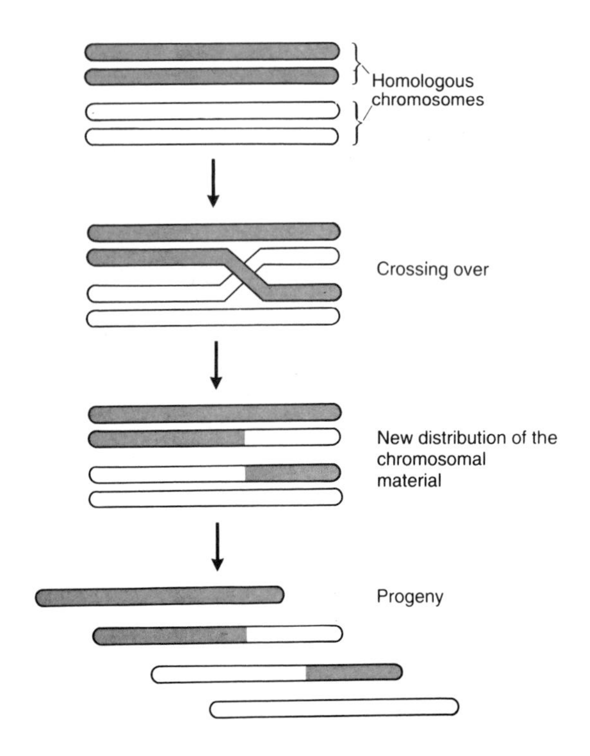

The general model (due to Robin Holliday[66]) describes an exchange of genes that takes place between a pair of homologous sequences, one from each parent. The exchange of homologous genes, which proceeds with minutely accurate genetic book-keeping, is restricted to the eucaryotes. For them, it is a cornerstone of genetic inheritance. To what extent the characteristic eucaryotic non-coding regions, the introns, may play a part in this is at present unknown.

We know much more about another kind of recombination, made use of by organisms that normally reproduce vegetatively. It has been investigated in great detail, not least because it has provided the basis for a new technology. The lowest level at which we encounter it is that of the conjugation of procaryotic, unicellular organisms, for whose evolution the facility for horizontal exchange of genes is also of crucial importance. This recombination is by no means as exact as the crossing-over principle of the eucaryotes, described above. Frequently the donor cell dies, because it has transferred essential genes to the receptor.

A vital function is fulfilled by mobile genetic storage units called *plasmids*. These are circular double helices of DNA containing between two thousand and several hundred thousand nucleotides. Although they are dispensable for the bacterium, the plasmids can become a part of its genetic apparatus and thereby transform it. They can also become integrated into the bacterial genome. Plasmids are able to endow bacteria with resistance toward an antibiotic.

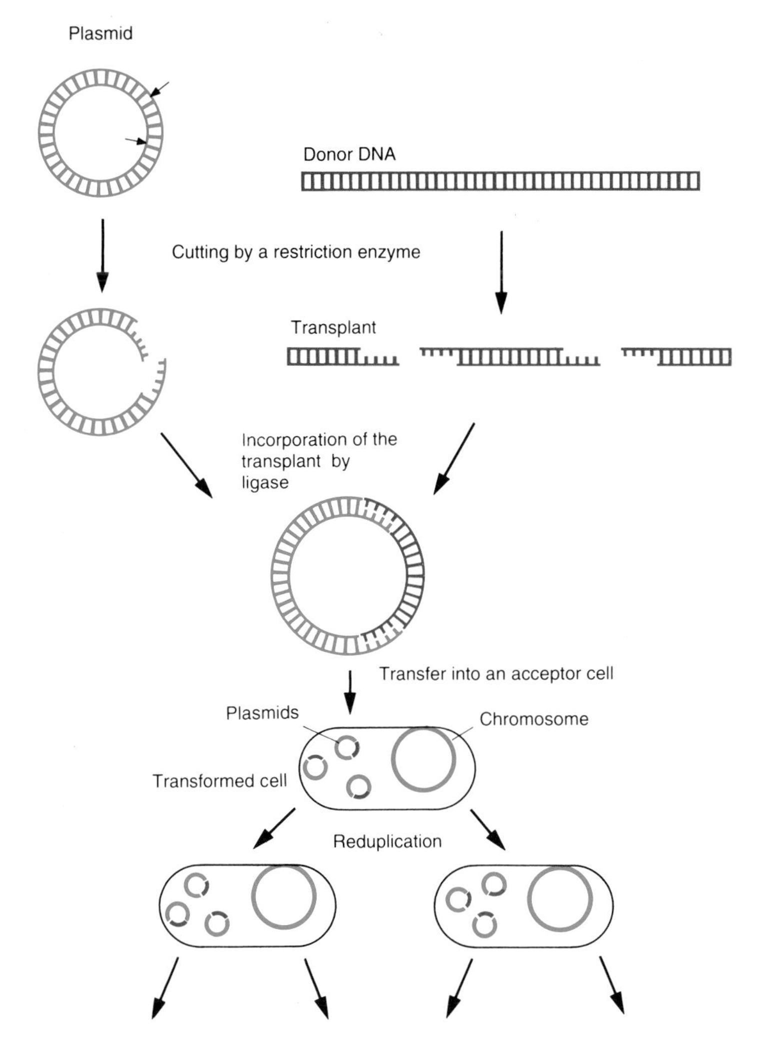

It is possible to insert genes into plasmids. For this reason, they are used in modern gene technology as packaging material for the transport of DNA, as are certain DNA phages. The plasmids transport the gene into a chosen host cell, where they are multiplied by the host cell's copying machinery. The gene introduced on the plasmid can likewise be copied, transcribed, and translated. In this way a valuable gene product, such as human insulin, can today be prepared industrially by microorganisms. The process that we have sketched is one of the foundations of modern biotechnology.

Indispensable tools in this process are the restriction enzymes, used to cut chromosomes at defined points, and the ligases, used to splice them together again. Cutting occurs at defined places because the enzyme recognizes a particular short sequence in the DNA (often a palindromic one); the enzyme then cuts across the strands in such a way as to leave loose ends that are complementary to one another. The acceptor strand is cut in the same way with the same enzyme, leaving loose ends of the same kind. By virtue of their complementarity, the loose ends from the donor DNA recognize the gap in the acceptor strand, and thus the foreign DNA can be spliced into the acceptor DNA at the right point. Apart from its technological importance, this procedure is an essential tool of research in molecular biology.

PART III

Résumé, Epilogue, Notes, Glossary

Résumé: Darwin is dead—long live Darwin!

Can the transition from non-living to living be understood and interpreted within our physico-chemical world picture? That is the fundamental question asked in this book.

The viruses stand at the borderline between living and inanimate matter. They display two properties of matter that are characteristic of the attribute 'life' and which, if we are to answer the question above, must be explained physically or chemically. The first of these properties is the complexity of all structures associated with life processes, from the proteins and the nucleic acids upwards; Erwin Schrödinger[67] called these 'aperiodic crystals'. The second is the way in which their molecular structures serve a clear functional purpose; Jacques Monod[68] spoke here of a 'teleonomy of organization'.

Many physicists believe that the currently accepted laws of physics are inadequate to explain these properties. According to Eugene Wigner,[69] even the phenomenon of reproduction is incompatible with the laws of quantum mechanics.

When we speak of reproduction in biology, we do not mean the exact reproduction of a physical state. This possibility was first investigated by Ludwig Boltzmann[70] as a problem in statistical mechanics, termed 'recurrence'. Reproduction in molecular biology is merely the replication of genetic information, the exact sequence of symbols in the nucleic acids, not the reproduction of the position and velocity of every atom. None the less, it is still perfectly reasonable to ask how a gene, the sequence of which is one out of 10^{600} (a one followed by six hundred zeros) possible alternatives of the same length, copies itself spontaneously and reproducibly. What is the physical cause of this?

There is no thermodynamic principle of the conservation of information. On the contrary, thermodynamics lays down the principle that entropy increases towards a maximum value at equilibrium, and this is related to the number of possible microscopic states that the system can assume. Thus, whenever a statistical fluctuation gives momentary rise to information, the information decays at once. This is the edict of the second law of thermodynamics. In a state of thermodynamic equilibrium, the spontaneous generation of information is impossible.

However, we do know that a living system, thanks to its metabolic processes, is far from being in a state of thermodynamic equilibrium. This is in particular the case for RNA and DNA molecules — for genes, which have been the especial objects of our attention. They are permanently subject to decay, and only by perpetual reproduction can they maintain their information content. But what is the nature of the physical forces that oppose the tendency towards equilibrium, and thus sustain an extremely improbable state? What physical laws allow the setting-up of an environment-related, teleonomic scale of values that define qualities such as 'right' or 'wrong', or 'good' or 'bad'? Can biology be reduced to physics at this decisive point? If it cannot, we are forced to postulate the existence of a *vis vitalis* — a demon at work outside of physical laws.

One such demon was conjured up by James Clerk Maxwell as early as in 1871, one endowed with the magic power of overcoming the second law of thermodynamics. Maxwell[71] considered a gas-filled vessel, divided by a partition into two chambers. In the partition, there was a tiny hole connecting the vessels and fitted with a trapdoor that could be opened or shut. The demon stood watch by the door. Whenever a rapidly moving molecule approached the door from the right-hand chamber, he opened it and allowed the molecule to pass through to the left-hand chamber; slow molecules, on the other hand, he held back. Molecules approaching from the left were sorted according to the opposite rule: fast ones were retained, while slow-moving ones were allowed to pass through to the right-hand chamber. The result of this manipulation was a thermal segregation, in which the left-hand chamber became hot and the right-hand one cold. This meant a decrease of entropy, contradicting the second law. In this century, Leo Szilard,[72] Dennis Gabor,[73] and Léon Brillouin[7] have used the idea of Maxwell's demon to illustrate the correlation between (negative) entropy and (absolute) information. In order to play his tricks, the demon needs information. This he pays for in the currency of entropy, which compensates for the entropy decrease in the vessel.

Monod[68] tried to invoke a similar demon, one that exploits random events. Szilard and Brillouin had shown that Maxwell's demon has to be fed, and the same applies to Monod's; however, this does not present a problem for biologists, since they know that life is in any case kept going by metabolism. Monod assumed that enzymes, in particular the self-regulating allosteric enzymes, take on the part of the Maxwell–Szilard–Brillouin demon and convert random fluctuations into information. Monod was clearly on the right path with this idea, as information can ultimately only emerge from non-information, that is, from random fluctuations. 'Enzymes function exactly in the manner of Maxwell's demon amended by Szilard and Brillouin, draining chemical potential into the processes chosen by the programme of which they are the executors.'[68]

However, the special, teleonomic structure of enzymes can hardly be responsible for the reproducible fixation of the (statistically) improbable state of a gene. Enzymes, in principle, also function in thermodynamic equilibrium, since they are able to catalyse reactions in both the forward and the reverse directions, without affecting the position of the equilibrium. They are themselves mere products of the selection into whose service they are pressed.

We should really be asking the questions: How did the programme that the enzymes execute come into being? How did enzymes become adapted to their teleonomic function? It is true that Monod regarded optimization as caused by evolution, which is connected to time's arrow and cannot run backwards (and thus takes place far from chemical equilibrium); and he saw that evolution, with the help of selection, 'draws from the inexhaustible well of chance'. But Monod went too far in believing that only chance can be a source of creation, while necessity, physical law, must be content with the minor role of a blind selection sieve. He does indeed feel a twinge of discomfort about this imbalance between chance and necessity: 'When one thinks about the tremendous journey of evolution over the past three thousand million years or so, the prodigious wealth of structures it has engendered, and the extraordinarily effective teleonomic performances of living beings, from bacteria to man, one may well wonder whether all this really can be the product of a vast lottery, in which natural selection has blindly picked the rare winners from among numbers drawn at utter random.'[68]

The demon with the special ability to transform fluctuations into information is not just some passive molecular 'rectifier' — he is a part of the mechanism of selection. Selection does not work blindly, and neither is it the blind sieve that, since Darwin, it has been assumed to be. Selection is more like a particularly subtle demon that has operated on the different steps up to life, and operates today at the different levels of life, with a set of highly original tricks. Above all, it is highly active, driven by an internal feedback mechanism that searches in a very discriminating manner for the best route to optimal performance, not because it possesses an inherent drive towards any predestined goal, but simply by virtue of its inherent non-linear mechanism, which gives the appearance of goal-directedness. That is the message of this book.

But what is new about this piece of insight? This brings us to the decisive questions. First, what is natural selection?

Far from equilibrium, and nowhere else, there operates a causal chain: replication → mutation → selection → evolution. This may not sound very new. Since Darwin, selection has generally been accepted to be the basis of evolution, and, since the foundation of quantitative population genetics by John B.S. Haldane, Ronald A. Fisher, and Sewell Wright in the first quarter of this century, reproduction and mutation have been accepted as the prerequisites of selection. What is new is the detailed structure, and also the consequences that we can draw from it. For one thing, molecular biology has provided us with much insight into the nature of genetic information and the principles according to which it is processed. For another, these insights have led to mathematical models that make precise predictions and can be tested by experiment.

Let us here consider one detail: the fact that replication implies autocatalysis. It can amplify a microscopic fluctuation until it is manifested macroscopically. But replication means more than *just* autocatalysis. The replica does not arise directly; it is produced via the intermediate product, the complementary negative template. For this reason, the early reproducing systems were forced to employ at least two symbol classes, and to use these, on average, equally often. From this row of chemical units there arose a readable sequence of symbols: information. From then on, such sequences were to encode the construction plans of all organisms, including humans. Chemistry had by now moved over to the side-stage; the biochemical principles governing the construction of human genes are largely the same as those for *E. coli*. Thanks to replication, genetic information has become immortal and, in spite of its perpetual destruction, it has continued to exist, in innumerable modified forms, for over three thousand million years. The vital element, after the origin of the first genes, has always been (adaptable) information. It represents a quality that far transcends chemistry. In Spiegelman's words: 'The nucleic acids invented human beings in order to be able to reproduce themselves even on the Moon.'

Replication that is forced to make use of various classes of monomer, each with its own set of interactions, also carries the potential for a further fundamental property: that of mutation. This is a simple result of errors in copying, caused by the fact that the energy of the interaction between two complementary nucleotides is not very much greater than the energy of thermal motion. So the occurrence of errors is inevitable, and does not call for any special mechanism. Quite on the contrary, the accumulation of genetic material during evolution has meant that errors must be restricted and where necessary reduced, and this has in fact been done, first by enzymes and later by sophisticated correction mechanisms.

Mutations are the source of evolutionary progress. For this reason, it is important that all mutants should have the ability to replicate. Thus, replication is an *inherent*, autocatalytic property of the whole molecular class. For the dynamic behaviour of a system, this means growth and competition between all the individuals that can replicate. The result of the competition — irrespective of whether the population as a whole grows or remains constant in number — is selection. This leads to an internal regulation of the relative numbers of the various mutants in the population. However, the result of the selection is not final, at least as long as the optimal state has not been reached. The choice, once made, is reviewed again and again, and is abandoned as soon as a better-adapted mutant appears on the scene, called into being by a statistical fluctuation. Mathematically, this means an instability in the solution of a differential equation; physically, a collapse of the previous distribution, which is replaced by a new one. The target of selection is the entire distribution, not just the wild type, which only in some cases is the best-adapted sequence within the populated ensemble that we call the quasi-species. Indeed, the wild type, which determines the selective decision, is often represented by a minority of the sequences actually present and frequently by a vanishingly small minority at that. It lies in the middle of the distribution and usually defines the consensus sequence. It is not necessarily the best-adapted individual sequence *per se*, since the target of selection is the quasi-species as a whole.

Particularly important are the neutral, or nearly neutral, mutants. These are mutants that are just as efficient as the wild type, or nearly so. They are now known to be present in far greater number than previously anticipated. Owing to their presence, the distribution of mutants around the wild type is far from symmetric: frequency in the population by no means decreases uniformly with increasing mutation distance from the wild type. Instead, mutants are concentrated in sequence space close to relatively well-adapted variants, and extend across sequence space over regions connecting these. In other words, even at a long distance from the wild type, reasonably well-adapted mutants are present in finite concentration, especially in connected regions of sequence space. This leads to a hierarchy of privilege among the mutants: not only does the selective advantage of well-adapted, almost neutral mutants lead to their better numerical representation, it also makes them arise more frequently. This is because a mutant at a relatively large distance from the wild type does not arise directly from the wild type; rather, it develops step by step from precursors at shorter distances from the wild type. The neighbours of well-adapted mutants are usually fairly well-adapted themselves, and are therefore also well represented. (This is connected with the fractal structure of the landscape, see Vignette 12.) Thus, quite at odds with the classical interpretation, the process of evolution is steered in the direction of the optimal value peak, and, thanks to the frequent criss-crossing of paths in multidimensional sequence space, this steering process is extraordinarily effective. The (quantitative) acceleration of evolution that this brings about is so great that it appears to the biologist as a surprising new *quality*, an apparent ability of selection to 'see ahead', something that would be viewed by classical Darwinians[74] as the purest heresy! The possibilities we have discussed of performing evolution experiments in the laboratory, and the success promised by evolutionary biotechnology, are both influenced by this qualitatively novel interpretation. It is based upon the ideas of sequence space and quasi-species, both elaborated in quantitative detail. These ideas fit effortlessly into the new physics of non-equilibrium states,[75] in which classical symmetries are broken. Hermann Haken has christened this modern branch of physics 'synergetics'.[76]

The new theme which this book has taken up is the detailed description of selection and evolution. Without this detail, the complexity and teleonomy of life would be incomprehensible. The old theme is and will remain Darwin's idea: the principle of evolution by natural selection.

The transition from inanimate to living structures took place with the increasing ability to wield information in a quasi-intelligent way. The first steps towards life were taken in a chemically rich environment, one in which information-storing molecular replicators appeared. Only these had the capacity for optimization, and thus for a teleonomic approach towards goal-determined behaviour. They preserve what has been attained, replacing it only with what is even more expedient. The genotype, the information contained in the nucleic acids, develops phenotypic semantics.

'Information' means primarily only information in the sense that a symbol sequence has a superior rate of replication, quality of replication, and life-span. It is these characteristics that, with the help of feedback (i.e. the endowment of the genotype with superior characteristics by its phenotype), give rise to the semantics of genetic information, which at the biological level depend ultimately upon the construction and the behaviour of the organism encoded in its symbol sequence.

The separation of genotype and phenotype began at the level of RNA molecules. The genotype was expressed in the physico-chemical properties of the molecule, the phenotype, that it encoded. The evolution of a translation system started out from the phenotypic characteristics of the participating RNA molecules. Ultimately, however, these were relegated to being the target structures of the protein executive that they encoded. The dichotomy of genotype and phenotype, manifested in material form as soon as translation appeared on the scene, could be bridged only by the establishment of a feedback loop that was superimposed upon the reproduction cycle and that integrated the translation products. A hypercyclic feedback loop of this kind made possible the coexistence of genotypes that initially were mutants of one and the same quasi-species, and it forced them to cooperate while still giving them the chance to differentiate. This effect was complemented by compartition, which was able to utilize selectively the functions of the phenotype for the benefit of the coding genotype. Before such a compartment became a cell, with an organized programme of division, many individual steps had to be integrated. The structural integration of all genes to give a genome, a giant DNA molecule, called for an apparatus of reproduction that worked with extreme precision, recognized and repaired errors, translated exactly, and could guarantee its own energy supply by metabolism. This necessitated a semipermeable demarcation from the world outside, that is, a cell membrane; it also required the regulation of the transport of substances through it. finally, the replication of the genome, the DNA, had to be synchronized with the duplication of the entire cell. Only then had the first level of life as we know it been reached.[77]

'What was life?' Thomas Mann asked this question insistently, and would not be satisfied with a facile answer. Climbing the steps towards life, from cell to man, has brought us to the point 'whence it sprang, where it kindled itself', and after which 'nothing happened arbitrarily, or without a cause'. But long and difficult will be our ascent from this lowest landing up to the topmost level of life, the level of self-awareness: our continued ascent from man to humanity.

Epilogue

The object of scientific research is that which is observable and inter-subjectively reproducible. Repeatable observations are set in relation to one another and fitted into a scheme that is free from contradictions. Striving towards scientific discovery is a collective activity of the mankind: nothing counts as proved unless it is accepted by the worldwide forum of science. This in no way excludes the possibility of error or misinterpretation. Yet no projection of reality is more reliable than the world-view supported by the edifice of scientific discovery.

Religious experience is based upon faith and thus possesses the independence that characterizes subjectivity. In this respect, it differs fundamentally from scientific knowledge. Owing to the limitations of human understanding, conflicts between religious and scientific world-views are hard to avoid. The frequently raised question 'Creation *or* evolution?' thus stems from a non-existent contradiction, since the word 'or' implies a confrontation between two incommensurable projections. I am well aware that these questions have occupied a central position in the subjective consciousness of many people. Nevertheless, although I am frequently asked about such matters, I have in this book left them untouched. Banal replies such as 'Evolution is the realization of creation by means of natural law' do little to satisfy, or to help, those who are seeking answers on this matter. They are in fact asking about something quite different, problems for which science offers no solution.

Notes and literature on the history of molecular biology

1. For a general introduction to palaeobiology, see M.G. Rutten, *The origin of life*, Elsevier, Amsterdam, 1971; H.K. Erben, *Die Entwicklung der Lebewesen*, Piper, Munich, 1985; Ross J. MacIntyre (ed.), *Molecular-evolutionary genetics*, Plenum, New York, 1985; M. Calvin, *Chemical evolution*, Oxford University Press, 1969.

2. For a critical review of the earliest assignments, see E.S. Barghoorn, The oldest fossils, *Scientific American*, May 1971. The most recent literature contains reports of the dating of fossils alleged to be up to four thousand million years old; however, these are not accepted by some.

3. The pioneer of sequence determination, both of proteins and of nucleic acids, is Frederick Sanger. A further important procedure for the sequence analysis of DNA was introduced by Allan Maxim and Walter Gilbert. Thanks to these rapid methods we now know a very great number of protein, RNA, and above all DNA sequences. These are stored in data banks such as the one at the European Molecular Biology Laboratory (EMBL) in Heidelberg or at the National Laboratories in Los Alamos, USA. They are also published in tables such as the *Atlas of protein sequence and structure* and the *Nucleic acid sequence data base*, both formerly issued by Margaret O. Dayhoff of the National Biomedical Research Foundation, Washington, DC, but now discontinued.

4. R. Winkler-Oswatitsch, M. Eigen, and A. Dress, Comparative sequence analysis — exemplified with tRNA and 5S-rRNA, *Chemica Scripta* **26B**, 59 (1986) and A statistical geometry in sequence space, *Proc. Natl. Acad. Sci. USA* **85**, 5913 (1988).

5. M. Eigen and R. Winkler-Oswatitsch, Transfer RNA: the early adaptor, *Naturwissenschaften* **68**, 217 (1981) and *Transfer RNA:* An early gene?, *Naturwissenschaften* **68**, 282 (1981); M. Eigen, B. Lindemann, R. Winkler-Oswatitsch, and C.H. Clarke, Pattern analysis of 5S-rRNA, *Proc. Natl. Acad. Sci. USA* **82**, 2437 (1985); M. Eigen, B.F. Lindemann, M. Tietze, R. Winkler-Oswatitsch, A. Dress, and A. von Haeseler, How old is the genetic code? Statistical geometry of tRNA provides an answer, *Science* **244**, 673 (1989).

6. Arguments for and against the hypothesis of 'directed panspermia' are discussed by F.H.C. Crick in *Life itself. Its origin and nature*, Simon and Schuster, New York, 1981.

7. For a summary of information theory and its physical implications, see L. Brillouin, *Science and information theory*, Academic Press, New York, 1962 (2nd edn, 4th imp, 1971).

8. M. Eigen and R. Winkler-Oswatitsch, *Das Spiel*, Piper, Munich, 1975; English translation, *Laws of the game*, Knopf, New York, 1981.

9. Darwin's famous work *The Origin of Species* has appeared in an edition with commentary by G.G. Simpson, based upon the 6th edn, which Darwin had revised himself and regarded as the final version. It is published by Collier-Macmillan (London, 1962). Its relationship to the work of Alfred Russel Wallace has been investigated by J. Langdon Brooks in *Just before the Origin*, Columbia University Press, New York, 1984.

10. The significance of Darwin's selection principle in modern biology is discussed in M. Eigen, Darwin and molecular biology, *Angew. Chem. Int. Ed.* **20**, 233 (1981).

11. For a non-specialist introduction to synergetics, see H. Haken, *Science of structure: synergetics*, Van Nostrand Reinhold, New York, 1984, or *Erfolgsgeheimnisse der Natur*, Deutsche Verlags Anstalt, Stuttgart, 1981.

12. The theory of the self-organization of biological macromolecules was developed by the author together with Peter Schuster and is described by M. Eigen and P. Schuster in *The hypercycle — a principle of natural self-organization*, Springer, Heidelberg, 1979; also by B.-O. Küppers in *Molecular theory of evolution*, Springer, Berlin, 1983 and in *Der Ursprung biologischer Information*, Piper, Munich, 1986 (English translation, *Information and the origin of life*, MIT Press, Cambridge, MA, 1990).

13. A detailed description of the quasi-species model can be found in a feature article by M. Eigen, J. McCaskill, and P. Schuster in *J. Phys. Chem.* **92**, 6881 (1988). The structure of an RNA quasi-species has been investigated experimentally by C.K. Biebricher, who used cloning and sequence analysis to study the distribution of mutants C.K. Biebricher, Darwinian selection of RNA molecules *in vitro*, in *Evolutionary biology*, edited by M.K. Hecht, B. Wallace, and G.T. Prance, Vol. 16, Plenum, New York, 1983.

14. Cloning experiments on bacteriophage Q_β have been carried out by C. Weissmann and his colleagues. See E. Domingo, D. Sabo, T. Tanaguchi, and C. Weissmann, Nucleotide sequence heterogeneity of an RNA phage population, *Cell* **13**, 735 (1978). Similar results have been obtained by E.

Domingo and co-workers for the virus of foot-and-mouth disease: see *Gene*, **11**, 333 (1980).

15. The error–threshold relation states the exact relationship between a quantity of information (number of nucleotides), the error rate, and the selective advantage of the dominant (wild-type) sequence. See M. Eigen, Self-organization of matter and the evolution of biological macromolecules, *Naturwissenschaften* **58**, 465 (1971). See also Notes 12 and 13.

16. The error–threshold relation has been tested on populations of various viruses. It was found that these natural populations operate just below this limit and are therefore very flexible in evolution. The Q_β phage is an example: E. Domingo, A. Flavell, and C. Weissmann, *In vitro* site-directed mutagenesis: generation and properties of an infectious extra-cistronic mutant of bacteriophage Q_β, *Gene* **1**, 3 (1976).

17. For a description of experiments for the analysis of the elementary steps in enzyme reactions, see M. Eigen, Die unmeßbar schnellen Reaktionen, in *Les Prix Nobel en 1967*, Stockholm, 1968, where further references can be found.

18. All classical evolution models are based upon the assumption of random mutations. See Notes 68 and 74.

19. M. Kimura and T. Ohta, *Theoretical aspects of population genetics*, Princeton University Press, 1971; M. Kimura, *The neutral theory of molecular evolution*, Cambridge University Press, 1983.

20. For a specialized account of this theory, see M. Eigen, Macromolecular evolution: dynamical ordering in sequence space, *Ber. Bunsenges. Phys. Chem.* **89**, 658 (1985); P. Schuster and K. Sigmund, Dynamics of evolutionary optimization, ibid., 668.

21. I. Rechenberg, *Evolutionsstrategie*, Problemata frommann-holzboog, Stuttgart–Bad Cannstatt, 1973.

22. For a summary, see S.L. Miller and L.E. Orgel, *The origins of life on earth*, Prentice-Hall, Eaglewood Cliffs, 1974; S.L. Miller in *Chemica Scripta* **26B**, 1986.

23. S.W. Fox and K. Dose, *Molecular evolution and the origin of life*, Freeman, San Francisco, 1972. A review of the work of S.W. Fox and his school can be found in Dose and Rauchfuss (Note 2).

24. See Note 22. Further articles by Orgel and his school (from 1974) have appeared in the journals *Nature*, *Science,* and in the *Journal of Molecular Evolution*. Summaries have been published by L.E. Orgel and by S.W. Miller in *Evolution of catalytic function*, Cold Spring Harbor Symposia in Quantitative Biology, 1987.

25. Articles by J. Oró on the synthesis of nucleic-acid bases are covered in the review references mentioned in notes 22 and 24. New, elegant syntheses have been developed by A. Eschenmoser at the Eidgenössische Technische Hochschule, Zürich, and a review article on this topic has been published in *Angew. Chem. Int. Ed.* **27**, 6 (1988).

26. D. Pörschke, in *Chemical relaxation in molecular biology*, edited by I. Pecht and R. Rigler, Springer, Heidelberg, 1977, p. 191; F.H.C. Crick *et al.*, *Origins of Life* **7**, 389 (1976). See also Notes 12 and 15.

27. This question has been pursued in recent experiments by L.E. Orgel and co-workers and by A. Eschenmoser (see e.g. *Nature* **310**, 602 (1984)). The experiments show clearly how much the replication of RNA and DNA depend upon the chiral nature of the monomers and the template strand, but they do not allow any definitive conclusions about the historical origin of chirality to be drawn.

28. The specialist literature contains many clues in support of the hypothesis that RNA preceded DNA in molecular evolution. These clues lie in the metabolism of RNA and DNA in present-day organisms, in the mechanism of DNA synthesis ('RNA priming'), and in the structural properties of RNA and DNA.

29. The book by M.G. Rutten quoted in Note 2 includes an extensive description of the development of our understanding of photosynthesis. Important contributions have been made by M. Calvin, *Chemical evolution,* Oxford University Press, 1969. The mechanism of photosynthesis has today been analysed almost completely down to its elementary steps. For a review see H.T. Witt, E. Schlodder, K. Brettel, and O. Saygin, Reaction sequences from light absorption to the cleavage of water in photosynthesis — routes, rates, intermediates, *Ber. Bunsenges. Phys. Chem.* **90**, 1015 (1986) (report of a conference on 'The Modelling of Complex Chemical Systems').

30. G. Eglinton and M. Calvin, Chemical fossils, *Scientific American* **216**, 32 (1967).

31. The molecular organization of the cell is treated in many excellent textbooks: J.D. Watson, Molecular biology of the gene, 3rd edn, Benjamin, Menlo Park, 1977; L. Stryer, *Biochemistry*, Freeman, San Francisco, 1975, 1981, 1988; A. Kornberg, *DNA Replication,* Freeman, San Francisco, 1974; E.L. Winnacker, *Gene und Klone*, Verlag Chemie, Weinheim, 1984 (English translation, *From genes to clones*, Verlag Chemie, Weinheim, 1987); J. Darnell, H. Lodish and D. Baltimore, *Molecular cell biology*, Scientific American Books, 1986.

32. Specialized reviews are given in *RNA phages,* edited by N.D. Zinder, Cold Spring Harbor Symposia in Quantitative Biology, 1975; M.H. Adams,

Bacteriophages, Interscience, New York, 1959; articles by C.K. Biebricher, E. Domingo, M. Eigen, J.J. Holland, P. Palese, and others in *RNA genetics,* (edited by E. Domingo, P. Ahlquist, and J.J. Holland), CRC Press, Baton Rouge, 1987. The first experimental demonstation of a natural hypercycle *in vivo* has been provided by M. Gebinoga (dissertation, Göttingen, 1990).

33. T.R. Cech, The generality of self-splicing RNA: relationships to nuclear mRNA splicing, *Cell* **44**, 207 (1986); W. Gilbert, The RNA World, *Nature* **618**, 319 (1986).

34. M. Eigen and P. Schuster, Stages of emerging life, *J. Mol. Evol.* **19**, 47 (1982); M. Eigen, W.C. Gardiner, P. Schuster, and R. Winkler-Oswatitsch, The origin of genetic information, *Scientific American* **244**, 88 (1981).

35. G.A. Dover, The spread and success of non-Darwinian novelties, in *Evolution processes and theory,* edited by S. Karlin and E. Nevo, Academic Press, New York, 1986.

36. H.G. Zachau, The human immunoglobulin X locus and some of its acrobatics, *Biol. Chem. Hoppe-Seyler* **371**, 1 (1990); S. Tonegawa, Somatic generation of antibody diversity, *Nature* **302**, 575 (1983); H.G. Zachau, M. Pech, H.-G. Klobeck, H.-D. Pohlenz, B. Straubinger, and F.G. Falkner, Wie entstehen die Antikörper?, 9th Fritz Lipmann Lecture, *Hoppe-Seylers physiol. Chem.* **365** 1363 (1984); S. Tonegawa, Nobel lecture in physiology and medicine, 1987, in *In Vitro Cellular and Developmental Biology*, 24th April, 1988.

37. Physico-chemical models for cell differentiation have been proposed for the freshwater organism *Hydra* by A. Gierer and H. Meinhardt and for the embryo of *Drosophila* by H. Meinhardt, and these have been compared with experimental results: see H. Meinhardt, Mechanisms of pattern formation during development of higher organisms: a hierarchical solution of a complex problem, *Ber. Bunsenges. Phys. Chem.* **59**, 691 (1985).

38. Mathematically based network models for the description of the development and function of the nervous system have been worked out by C. von der Malsburg: see Nervous structures with dynamical links, *Ber. Bunsenges. Phys. Chem.* **59**, 703 (1985). Both the models of von der Malsburg and those of Meinhardt (Note 37) show mathematical parallels with models of evolution.

39. M. Eigen, *New concepts for dealing with the evolution of nucleic acids*, Cold Spring Harbor Symposia in Quantitative Biology, Vol. 52, p. 307, 1987. See also the references in Note 20.

40. Known sequences of transfer RNA molecules have been collated by M. Sprintzl, J. Weber, J. Blank, and R. Zeidler, *Nucleic Acids Research* **17**, r1

(1989). For a collation of known rRNA sequences, see Th. Specht, J. Wolters, and V.A. Erdmann, *Nucleic Acids Research* **18**, r2215 (1990).

41. P. Palese, Rapid evolution of human influenza virus, in *Evolutionary processes and theory* , edited by S. Karlin and E. Nevo, Academic Press, New York, 1985.

42. The construction of a phylogenetic tree is a problem of optimization that can be solved in different ways. The school of E. Margoliash has pioneered this. An important approach is the 'method of maximum parsimony': W.M. Fitch, *J. Mol. Biol.* **16**, 9 (1966). The phylogenetic tree of the procaryotes comes from tRNA and 5S-rRNA sequence data of H. Hori, C. Woese, and H. Küntzel (compare Notes 4, 39, and 40).

43. In the early 1950s these compounds had been both structurally characterized and synthesized: A.R. Todd, *Les Prix Nobel*, Stockholm, 1957.

44. The classical papers by J.D. Watson and F.H.C. Crick on the structure of the nucleic acids appeared in 1953: A structure for deoxyribose nucleic acid, *Nature* (25 April 1953), p. 737 and Genetical implications of the structure of deoxyribonucleic acid, Nature (30 May 1953), p. 964. *The double helix*, Watson's popular account of the discovery, became a best seller (Weidenfeld and Nicolson, London, 1968). See also H.F. Judson's *The eighth day of creation*, Simon and Schuster, New York, 1979. A documentation of the subsequent rapid development of molecular biology, from the discovery of this structure to the cloning of genes, has been made by J.D. Watson and J. Tooze, *The DNA story* Freeman, San Francisco, 1981. Milestones in this development have been the discovery and isolation in 1958 of the enzyme that copies DNA (DNA polymerase) by A. Kornberg and of the enzyme that transcribes DNA to RNA in 1960 by J. Hurwitz and S. Weiss; the idea of an RNA 'messenger' (mRNA) and the discovery of the mechanism of gene regulation by F. Jacob and J. Monod in 1960–1; the demonstration of the existence of mRNA by S. Brenner, F. Jacob, and M. Meselson in 1961 and its confirmation by S. Spiegelman using DNA–RNA hybridization in 1961; the solving of the genetic code between 1961 and 1966 (see Note 50); the discovery of plasmids in 1965 by J. Lederberg; and, finally, a decisive step towards gene technology, the discovery of the restriction enzymes in 1970 by W. Arber and H. Smith and the cataloguing of restriction sites by D. Nathans.

45. The structures of the nucleic acids and the methods for determining them have been summarized in the monograph by W. Sänger, *Principles of nucleic acid structure*, Springer, Heidelberg, 1984.

46. The structure of DNA found by Watson and Crick in 1953 is today known as the B form. It has been joined by further structures, especially the A and

Z forms. The last is twisted in a sense opposite to that of the other two. Its structure was first seen directly by A. Rich and co-workers, by the use of X-ray diffraction. Previously, F. Pohl and T. Jovin had concluded from measurements of optical polarization that such a structure existed, and had shown under what external conditions the structure could be induced to appear and disappear. It is possible that Z-DNA plays a part in the regulation of information accessing.

47. The sequence (primary structure) and hydrogen-bonding pattern (secondary structure) of a tRNA were first determined by R.W. Holley (*Science* **147**, 1462 (1965)), followed later by observations on two others by H.-G. Zachau (*Angew. Chem.e Int. Ed.* **5**, 422 (1966). In those days the determination of the sequence of an RNA took years; today it can take weeks or less. Not until the early 1970s was the folding in space (tertiary structure) of a tRNA first discovered, by X-ray crystallography, in the laboratories of A. Rich and S.-H. Kim (S.H. Kim, G.J. Quigley, F.L. Suddath, A. McPherson, D. Sueden, J.J. Kim. J. Weinzierl, and A. Rich, *Science* **179**, 285 (1973)) and A. Klug (*J. Mol. Biol.* **89**, 511 (1974); ibid. **108**, 619 (1976)). Until now, all tRNA molecules investigated show a high degree of congruence in their primary, secondary, and tertiary structures.

48. R.E. Dickerson, *Scientific American* **226**, 58 (1972); see also R.E. Dickerson and I. Geis, *The structure and action of proteins*, Harper and Row, London, 1969.

49. The pioneers of the determination of protein structures by X-ray crystallography were M.F. Perutz and J.C. Kendrew. The determination of the spatial structures of the molecules of myoglobin and haemoglobin in the 1950s are milestones in the development of molecular biology (see *Les Prix Nobel*, Stockholm, 1962). The method is described more extensively in the books by Stryer (Note 31) and by Dickerson and Geis (Note 48).

50. The genetic code was 'broken' principally in the laboratories of M.W. Nirenberg and H. G. Khorana. The starting point for this was the availability of a homogeneous mRNA strand, in this case poly(U), an RNA made up of U polymers alone. It cannot encode any other triplet 'word' than UUU. This polymer was synthesized with the help of an enzyme discovered by Severo Ochoa and Marianne Grunberg-Manago. If this polymer was presented with a cell extract capable of carrying out protein biosynthesis, it began to produce a uniform peptide chain, a polymer made up from the monomer phenylalanine only (J.H. Matthaei and M.W. Nirenberg, *Proc. Natl. Acad. Sci. USA* **47**, 1580 (1961)). The deciphering of all the complex combinations of the code is due to the work of the brilliant synthetic chemist H.G. Khorana, who prepared synthetic nucleotide strands with all the possible repetitive code combinations (*Les Prix Nobel*, Stockholm, 1968). Decisive steps that led to an understanding

of the process of translation were F.H.C. Crick's inspired and far-sighted hypothesis of adaptors that put into effect the assignment of amino acids to their codons (see Note 47) and C. Yanofsky's proof of the collinearity of gene and protein structure, found in its simplest form in the procaryotes. Already in the work of Nirenberg indications can be found suggesting the universality of the genetic code.

51. R. Dulbecco showed around 1960 that the infection of a cell with DNA tumour viruses (SV40) results in the viral DNA becoming integrated into that of the cell, and that the transformation of the cell into a tumour cell depends upon this change in its genetic material. In 1964, D. Baltimore and H. Temin discovered independently of one another the enzyme that rewrites viral RNA as DNA, and called it reverse transcriptase. The existence of this enzyme provided the proof of Temin's 'provirus' hypothesis. After 1970, the mechanism of the reverse transcription was solved, largely by Baltimore, and the process of transformation thereby reduced to the mechanism which Dulbecco had postulated. Although until 1975 no human cancer had been shown to be connected directly to the action of retroviruses, it very soon became clear that the principle is of general relevance for the formation of tumours (*Les Prix Nobel*, Stockholm, 1975). See also Vignette 13.

52. The reproductive enzyme of the bacteriophage Q_β is able to make RNA *de novo*, that is, in the absence of any template strands, as first described by M. Sumper and R. Luce, *Proc. Natl. Acad. Sci. USA* **72**, 162 (1975). (Template strands usually possess a specific recognition sequence.) C. Biebricher has shown in many experiments that the homogeneous sequences that arise, 100–200 nucleotides in length, are not copies of trace RNA molecules present as impurities accompanying the enzyme, but instead are genuine products *de novo*. Their distributions behave like quasi-species and adapt themselves with a high degree of flexibility in response to changes in their environment (C.K. Biebricher, M. Eigen, and R. Luce: *J. Mol. Biol.* **149**, 369 and 391 (1981); *Nature* **321**, 89 (1986)). Biebricher was able to clone individual mutants in a quasi-species distribution of this kind and, using reverse transcriptase, to rewrite their sequences as DNA, and thus to sequence them. This was the first quantitative measurement of the behaviour of a quasi-species (see note 13). RNA sequences synthesized *de novo* are of especial value for evolution experiments.

53. In addition to the experiments described in Notes 14 and 54, reference should be made to the investigations into the vesicular stomatitis virus in the laboratory of J.J. Holland (*Science* **215**, 1577 (1982)).

54. The mutation rates for the influenza A virus and the polio virus, both of which possess a single-stranded RNA genome, have been investigated in

the laboratory of P. Palese by the methods of cloning and sequence analysis. For the influenza A virus, the high error rates typical of the RNA viruses were found. However, for the polio virus, no mutants were found under the experimental conditions chosen. Palese's method required the multiplication of the number of viruses up to a thousand million copies, the minimum number with which a sequence can be determined. During this procedure, which takes nearly two days, there arise many mutants that are neutral with respect to the wild type. Strong selection pressure, allowing only very few alternatives to the wild type, would in this experiment sharply reduce the number of genuinely neutral mutants. It therefore seems likely that the mutation rate is as high for the polio virus as for the influenza virus, and it is the pressure towards adaptation that differs between the two species. The difficulty mentioned in Vignette 9 refers to this fact. C. Weissmann's method of determining the mutation rate by measuring the number of revertants (Note 14) is not affected by this problem.

55. See J. Swetina and P. Schuster, Self-replication with error — a model for polynucleotide replication, *Biophys. Chem.* **16**, 329 (1982).

56. See M. Eigen and C.K. Biebricher, Sequence space and quasi-species distribution in RNA Genetics II (see Note 32).

57. See D.R. Mills, R.L. Peterson, and S. Spiegelman, An extracellular Darwinian experiment with a self-duplicating nucleic acid molecule, *Proc. Natl. Acad. Sci. USA* **58**, 217 (1967). Later experiments were aimed at adapting the RNA molecules to particular inhibitors of replication (R. Safhill, H. Schneider-Bernloehr, L.E. Orgel, and S. Spiegelman, In vitro selection of bacteriophage Q_β variants resistant to ethidium bromide, *J. Mol. Biol.* **51**, 531 (1970); D.R. Mills, F.R. Kramer, and S. Spiegelman, Complete nucleotide sequence of a replicating RNA molecule, *Science* **180**, 916 (1973)). Spiegelman obtained in three successive steps mutants containing respectively one, two and three errors; the last grew twice as fast as the wild type in the presence of the growth-inhibiting substance. By mutation of his *de novo* products, M. Sumper (Note 52) obtained mutants with an orders-of-magnitude greater adaptability to their environment. One of his variants could even be made to grow in the presence of the enzyme RNase T_1, which normally breaks down RNA. This would destroy the wild type completely. However, the adapted variant replicated in the presence of ribonuclease as well as the wild type did under normal conditions.

58. See Y. Husimi and H.C. Keweloh, Continuous culture of bacteriophage Q_β using a cellstat with a bubble-wall growth scraper, *Rev. Sci. Instrum.* **58**, 1109 (1987), and also Y. Husimi, K. Nishigater, Y. Kinoshita, and T. Tanaka, *Rev. Sci. Instrum.* **53**, 517 (1982).

59. The experimental curves shown are taken from the diploma thesis of A. Schwienhorst (Göttingen and Münster, 1987).

60. The apparatus shown was constructed as part of the doctorate in measurement and regulation engineering of H. Otten (Göttingen and Brunswick, 1987). The evolution studies were carried out by G. Bauer as part of a doctorate in biochemistry (Göttingen and Brunswick, 1990).

61. R.W. Hamming, *Coding and information theory*, Prentice-Hall, Englewood Cliffs, 1980.

62. B.B. Mandelbrot, *The fractal geometry of nature*, Freeman, New York, 1983. See also H.O. Peitgen and P. Richter, *The beauty of fractals*, Springer, Berlin, 1986.

63. The T phages have played a central role as model systems in the development of molecular genetics. The pioneers of this development were M. Delbrück and S.E. Luria. (See P. Fischer's *Licht und Leben. Ein Bericht über Max Delbrück, den Wegbereiter der Molekularbiologie*, Universitätsverlag Konstanz, 1985).

64. A compartment model has been simulated by C. Bresch, U. Niesert and D. Harnasch (Hypercycles, parasites and packages, *J. Theor. Biol.* **85**, 399 (1980); also U. Niesert, D. Harnasch and C. Bresch, Origins of life between scylla and charybdis, *J. Mol. Evol.* **17**, 348 (1981)). The model shows that a relatively high accuracy of replication, in comparison with the quantity of information, is needed in order to maintain the stability of the system. However, it ignores the presence of mutants with descending degrees of fitness. Yet this is of critical importance for determining the height of the error threshold. If only the wild type has a finite fitness (selection value), then selection can always take place, since there are no real competitors. On this point, see M. Eigen, W. Gardiner, and P. Schuster, Hypercycles and compartments. Compartments assist — but do not replace — hypercyclic organisation of early genetic information, *J. Theor. Biol.* **85**, 407 (1980).

65. Especially relevant for the subject of recombinant DNA is the work by Arber, Smith, and Nathans alluded to in Note 44 (*Les Prix Nobel*, Stockholm, 1978), as well as the books by Winnacker, Watson, and Tooze and by Darnall, Lodish, and Baltimore (Note 31). The mechanism of genetic recombination is treated in greater detail in the textbooks by Watson and Stryer (Note 31).

66. For a specialized review, see H. Potter and D. Dressler, *In vitro* system from *E. coli* that catalyses generalized genetic recombinations, *Proc. Natl. Acad. Sci. USA* **75**, 3698 (1978) and F. Stahl, *Genetic recombination*, Freeman, San Francisco, 1979.

67. E. Schrödinger, *What is life?*, Cambridge University Press, 1944.

68. J. Monod, *Le hasard et la nécessité*, Éditions du Seuil, Paris, 1970;

English translation, *Chance and necessity*, Knopf, New York, 1971, and Collins, London, 1972. Quotations from the English translation with minor amendments.

69. See E. Wigner, in *The logic of personal knowledge*, edited by E. Shils, Free Press, Glencoe, 1961 (Festschrift for the seventieth birthday of Michael Polanyi, 11 March 1961).

70. The time taken for the reproduction of two microstates is called the 'Poincaré recurrence time'. L. Boltzmann (*Ann. Phys.* **57**, 773 (1896); ibid. **60**, 392 (1897)) estimated this time for an ideal gas. He considered 10^{18} molecules, occupying one cubic centimetre and having an average speed of five hundred metres per second. He calculated that it would take more than $10^{10^{19}}$ (ten to the power of ten to the nineteenth) years for all of these molecules to adopt positions within 10^{-7} centimetres and velocities within one metre per second of their starting values.

71. J.C. Maxwell *Theory of heat* (p. 328); see also J.H. Jeans *Dynamical theory of gases*, 3rd edn, p. 183, Cambridge University Press, New York, 1921. (The latter work, available through libraries, contains an extensive description of Maxwell's ideas, including many direct quotations.)

72. See L. Szilard, Über die Entropieverminderung in einem thermodynamischen System bei Eingriffen intelligenter Wesen, *Z. Phys* **53**, 840 (1929).

73. See D. Gabor, *MIT lectures*, 1951 (discussed in detail by L. Brillouin, see Note 7).

74. See F.J. Ayala and J.A. Kiger, *Modern genetics*, 2nd edn, Benjamin/Cummings, Menlo Park, 1984. A classic text in population genetics is S. Wright's *Evolution and the genetics of populations*, University of Chicago Press. See also E. Mayr's *The nature of Darwinian revolution, Science,* **176**, 981 (1972) and *Evolution und die Vielfalt des Lebens*, Springer, Berlin, 1979.

75. For a specialized treatise, see P. Glansdorff and I. Prigogine, *Thermodynamic theory of structure, stability and fluctuations*, Wiley-Interscience, London, 1971. This includes a description of symmetry breakdown and the instability that results. For a simpler treatment, see I. Prigogine, *From being to becoming — time and complexity in the physical sciences*, Freeman, San Francisco, 1980, and I. Prigogine and I. Stengers, *Order out of chaos*, Bantam Books, New York, 1983, and Heinemann, London, 1984.

76. See H. Haken, *Synergetics. An introduction*, Springer, Heidelberg, 1977, 1983.

77. In addition to the various works already cited, the following are of interest in connection with the restricted problems of molecular evolution.
A. Bablyoantz, *Molecules, dynamics and life: an introduction to self-organization of matter*, Wiley-Interscience, New York, 1986.
A.G. Cairns-Smith, *Genetic takeover — and the mineral origins of life*, Cambridge University Press, 1982.
F.H.C. Crick, *Of molecules and men*, University of Washington Press, Seattle, 1966.
F. Dyson, *Origins of life*, Cambridge University Press, 1985.
F. Jacob, *La Logique du vivant — une histoire de l'hérédité*, Gallimard, Paris, 1970 (English translation *The logic of life*, Panthéon, Paris, 1982, and Random, New York, 1982).
F. Jacob, *The possible and the actual*, Pantheon, New York, 1982.
H.-D. Försterling and H. Kuhn, *Moleküle und Molekülanhäufungen*, Springer, Berlin, 1983, Chap. 21; H. Kuhn, Self-organization of matter and the evolution of early forms of life, in Biophysics, (edited by W. Hoppe, W. Lohmann, H. Markl, and H. Ziegler, Springer, Berlin, 1983.
B.-O. Küppers, *Information and the origin of life* (see Note 12).
B.-O. Küppers (ed.), *Leben = Physik + Chemie? Das Lebendige aus der Sicht bedeutender Physiker. Ein Lesebuch*, Piper, Munich, 1987.
B.-O. Küppers (ed.), *Ordnung aus dem Chaos*, Piper, Munich, 1987.
G.S. Kutter, *The universe and life — origins and evolution*, Jones and Bartlett, Boston/Portola Valley, 1987.
S. Lifson, Chemical selection, diversity, teleonomy and the second law of thermodynamics. Reflections on Eigen's theory of self-organization of matter, *Biophys. Chem.* **26**, 303 (1987).
A. Lwoff, *Biological order*, MIT Press, Cambridge, MA, 1962.
L. Orgel, *The origins of life*, Wiley, New York, 1973.
A.I. Oparin, *Genesis and evolutionary development of life*, Academic Press, New York, 1968.
P. Schuster and K. Sigmund, From biological macromolecules to protocells — the principle of early evolution, in *Biophysics*, edited by W. Hoppe, W. Lohmann, H. Markl, and H. Ziegler, Springer, Berlin, 1983.
J. Maynard-Smith, *The problems of biology*, Oxford University Press, 1986.
E. Schoffeniels, *L'anti-hasard*, Gauthier-Villars Éditeur, Paris, Bruxelles, Montreal, 1943; English translation *Anti-chance*, Pergamon, Oxford, 1976.
A. Unsöld, *Evolution kosmischer, biologischer und geistiger Strukturen*, Wissenschaftliche Verlagsgesellschaft, Stuttgart, 2nd edn, 1983.
Dynamically organized systems, *Ber. Bunsenges. Phys. Chem.* **89**, 563 (1985).
Molecular evolution of life, edited by H. Baltscheffsky, H. Jörnvall, and R. Rigler, published on behalf of the Royal Swedish Academy by Cambridge University Press and *Chemica Scripta*, 1986.

Evolution of catalytic function, Cold Spring Harbor Symposia in Quantitative Biology Vol. 52, 1987.
Life, origin and evolution, readings from *Scientific American,* Freeman, San Francisco, 1979.

Glossary

For a more detailed explanation of these and other terms, the reader may consult M. Abercrombie, M. Hickman, M.L. Johnson, and M. Thain, *The New Penguin Dictionary of Biology*, Penguin Books, London, 1990 (8th edn) and J.M. Lackie and J.A.T. Dow, *A Dictionary of Cell Biology*, Academic Press, London, 1989.

acceptor strand
: A strand of nucleic acid into which a piece of foreign genetic material can be or has been placed.

achiral
: A molecule that does not display chirality* (has no screw-like symmetry) is termed achiral.

active site or *active centre*
: The region of an enzyme* molecule in which the substrate is bound and chemically transformed.

adenine (A)
: One of the bases of the nucleic acids; see *base*.

adenosine triphosphate (ATP)
: An energy-rich phosphate ester; important as a molecular energy store in the cell. The splitting of one of the bonds linking the phosphate groups releases the energy and makes it available for biochemical reactions.

aerobic
: A term used to denote organisms (aerobes) that live in the presence of atmospheric oxygen, and the metabolic processes that take place in them.

AIDS
: Acquired immune deficiency syndrome: a viral disease that paralyses the immune* system.

alanine
: One of the twenty common naturally occurring amino* acids; a basic building block of the proteins.

* *quod vide*

allosteric enzyme

An enzyme* whose biological properties are changed when a small molecule (an *effector*) becomes bound to a site other than the active* centre. The binding of the effector generally alters the conformation of the enzyme.

alpha-helix

One of the structures adopted by the polypeptide* chains of proteins; a helical (spiral-staircase) structure stabilized by hydrogen* bonds between each CO group and the fourth-nearest NH group.

amino acid

Many amino acids occur in Nature. The twenty commonly occurring ones are the building blocks of the proteins. These twenty are all L-stereoisomers* (left-handed) and they all have the same basic structure, differing only in their single, functional side chain. See Vignette 6.

amino-acid sequence

See *primary structure*.

aminoacyl synthetase

One of at least twenty different enzymes* that, with the help of ATP*, generate an activated amino* acid and then bind it to the tRNA* adaptor that is specific for the amino acid in question.

anaerobic

A term used to denote organisms (anaerobes) that live in the absence of atmospheric oxygen, and the metabolic processes that take place in them.

Ångström unit

A measure of length used to specify sizes on the atomic scale. One Ångström unit is equal to a ten thousand millionth of a metre (10^{-10} m).

annealing

A heat treatment applied to metal, glass, plastic and, in biochemistry, nucleic acids. The treatment consists in raising the temperature slowly, in the region of the transition, and then lowering it slowly. In this way, tension and heterogeneity are removed. For nucleic acids, annealing promotes the optimal joining of complementary* regions to give double* helices.

antibiotic

A small molecule that at low concentration specifically inhibits the growth of microorganisms or kills them.

antibody

A protein produced in the bodies of vertebrates as a defence against foreign substances (antigens*), which are bound and, with the help of subsequent mechanisms of destruction, rendered harmless.

anticodon
A sequence of three nucleotides* (bases) in tRNA*. During the translation* process the anticodon interacts with the codon* (which also consists of three nucleotides) and is thus responsible for the incorporation of the correct amino* acid into the growing protein chain.

antigen
A macromolecular* substance that, when introduced into a vertebrate organism from outside, causes the production of antibodies*; these bind the foreign matter and render it harmless.

archaebacteria
Also called metabacteria. Special species of procaryotic* microorganism. Archæbacteria are usually found adapted to extreme (presumably 'archaic') environmental conditions, such as high temperature or high salt or sulphur content. They also generate unusual metabolic products, such as methane.

aspartic acid
One of the twenty common naturally occurring amino* acids; a basic building block of the proteins.

autocatalysis
Autocatalysis takes place when the product of a reaction is also a catalyst* for the same reaction.

axiom
A self-evident principle that does not require proof and that cannot be justified or further reduced logically.

bacteriophage or *bacteriovirus*
Viruses that multiply in bacteria.

base
In biochemistry, the basic components of the nucleic acids. These include principally adenine (A), cytosine (C), guanine (G), and thymine (T) with its close relation uracil (U). They are the four symbols of genetic information.

base pairing
A specific interaction between bases*. This occurs between guanine and cytosine, and between adenine and thymine/uracil.

binary
Made up of two units. In logic, a choice between two and only two possibilities.

biosynthesis
The formation of organic compounds and cellular components in the living organism, with the help of catalysis* by the appropriate cellular components

(e.g., enzymes*). Biosynthesis can be made to occur *in vitro* with the help of isolated cell components.

bit
Short for 'binary digit'. Any unit of information corresponding to an 'either–or' decision, such as 0 and 1, or R and Y.

blue alga
A cyanobacterium*, belonging to the kingdom of the eubacteria*.

Brownian motion
Collisions with molecules undergoing thermal movement results in a random, chaotic motion of small particles suspended in a liquid. They were first described by their discoverer, the English botanist Robert Brown. The term 'Brownian' has also been applied by Benoit Mandelbrot to the profile generated by a diffusion* process.

bursting
The disruption of a cell that has been infected by a virus, accompanied by the emergence of viral particles.

catalysis, catalyst
The acceleration of a chemical reaction. The catalyst (the agent of catalysis) can be a molecule, an area on a metal surface or a molecular complex. Most biochemical calatysts are proteins, and are termed enzymes*.

cell differentiation
The process by which cells in a multicellular organism acquire and retain their specialized function.

cell line
A population of cells, all of which have arisen by repeated cell division starting from a single ancestor cell and are its direct genetic descendants.

cellstat
A flow reactor for the raising and the adaptation of viruses that are kept supplied with a constant population of host cells (from a turbidostat*).

chemostat
An apparatus in which a bacterial culture can grow continuously.

chirality
The 'handedness', for example of a screw or a helix (right- or left-handed). A molecule exhibiting chirality is termed 'chiral'.

chloroplast
The particles responsible for photosynthesis*, coloured green by the chlorophyll they contain. Common to the algae and higher plants.

chromosome
Unit of heredity, containing many genes*; a linear structure in which the hereditary material is contained.

clone
A set of genetically identical descendant cells that have resulted from the asexual replication of a single ancestor cell.

cloning
The production of clones* of a given cell, by gene-technological or other methods. Also applied to the multiplication of DNA molecules in a plasmid*.

coat protein
A protein from which the coat of a virus is built up.

codon
A sequence of three consecutive nucleotides* that code for an amino* acid or for the signals 'start' or 'stop'.

coli bacterium
See *Escherichia coli*.

collinear, co-linear
Retaining a linear pattern. For example, the amino-acid sequence of a protein is collinear with the sequence of codons* in its messenger RNA*.

complementary, complementarity
(a) In physics, complementarity is the characterisation of a system in different, independent ways, that complement each other. (b) For nucleic acids, it is the ability to undergo specific base-pairing* interactions (G with C, and A with T or U). This underlies the mechanism of complementary copying (see Vignettes 4 and 5).

condensation
(a) The conversion of a gas or vapour to a liquid, by compression or by cooling. (b) A chemical reaction in which two molecules expel a simple molecule (such as water) and combine to give a single large molecule.

conjugation
In bacterial genetics, the transitory coupling of two bacterial cells, during which a piece of DNA is transferred from a donor to an acceptor cell.

consensus sequence
In a set of homologous* sequences, the consensus sequence is the one that results from taking the most commonly occurring symbol at each position.

crossing-over
The process of exchange of genetic material between two chromosomes*. A fundamental part of sexual reproduction.

cyanobacterium
A micro-organism that, while possessing the procaryotic* cell structure typical of bacteria, carries out photosynthesis*, liberating oxygen, in the manner typical of the chloroplasts* of the (eucaryotic*) plants.

cytochrome
Important respiratory enzymes*, of which the best studied is cytochrome *c*.

cytoplasm
The region of a cell outside the nuclear membrane, separated from its environment by the cell wall. It can thus be taken to mean the entire contents of the cell except the nucleus, or, in a more narrow sense, the entire protoplasm, that is, the cell contents without the organelles and the cell skeleton.

cytosine (C)
One of the bases of the nucleic acids; see *base*.

de-repression
The reversal of repression*, that is, the activation of a previously repressed gene* by the removal of a bound repressor* molecule.

deletion
(a) The omission of symbols in the copying process. (b) The loss of a part of the genetic material from the chromosome*. The amount of deleted material can be anything from a single nucleotide* to a segment containing several genes.

deoxyribonucleic acid
See *DNA*.

deoxyribose
The sugar component of the nucleotides* from which deoxyribonucleic acid (DNA) is assembled. Related to ribose* by the removal of one oxygen atom.

dichotomy
Splitting into two. In biology, the division of a genus into two species; in logic, the subdivision of a category into two sub-categories showing opposite behaviour.

diffusion
Transport by thermal molecular movement. A mechanism for the transport of molecules and atoms from regions of high concentration to regions of lower concentration.

divergence
In the context of this book, the departure of lines of development of sequences caused by the accumulation of mutations*.

DNA
Deoxyribonucleic acid; the molecular carrier of hereditary information. Usually occurs as a double-stranded, double-helical* structure.

DNA polymerase
A polymerase* that catalyses* the stepwise synthesis (polymerization) of DNA strands in the complementary* copying of DNA.

donor strand
A strand of DNA from which a gene* sequence is removed and transferred to an acceptor strand*.

double helix
Two helically intertwined strands of DNA (see Vignettes 4 and 5).

drift, genetic
Progressive change in a gene* due to neutral* mutations, that is, those that are neither advantageous nor disadvantageous in selection.

dynamic theory
A physical theory of moving objects or of systems that alter in time.

electron donor, electron acceptor
In redox (reduction–oxidation*) reactions, an electron is transferred from an electron donor (reducing agent) to an electron acceptor (oxidising agent). A biological example is the transport of electrons along the respiratory chain* by the cytochromes*.

enantiomer
Enantiomers are isomers* whose only difference is that they are of opposite chirality* (left- or right-handed).

endocytosis
Integration of a substance into a cell by uptake into and release from its membrane.

energy-rich phosphate bond
See *adenosine triphosphate*.

entropy
A thermodynamic* function of state. Entropy is complementary to temperature. The entropy of a system tells us how many detailed states, or degrees of freedom, are accessible to the average kinetic energy of thermal motion expressed in the temperature. The product of temperature and entropy has the dimensions of energy. In information theory*, the concept of (negative) entropy is used as a measure of information needed: it is the average number of yes/no decisions (bits*) required to identify a message of given length. As far as order can be taken to mean a restriction in the number

of possible states, an increase in entropy can be regarded as a decrease in order.

enzyme
A biological catalyst*, usually a protein molecule or a complex of protein molecules. An enzyme has a particular affinity (*specificity*) for the molecule (the *substrate*) that is to be transformed chemically.

error threshold
A critical value of the mutation* rate, above which errors accumulate and soon lead to the complete loss of information (the *error catastrophe*). Stable selection requires that the error rate lie below the error threshold.

Escherichia coli or *E. coli*
A species of bacterium. It occurs in the gut (the *gut flora*) of warm-blooded animals.

eubacteria
A recent term for the kingdom of bacteria, which includes most of the procaryotes* (such as *E. coli, Streptomyces*, phototropic bacteria and cyanobacteria*). The remaining procaryotes* are archaebacteria*.

eucaryote, eucaryotic
An organism distinguished by its well-formed nucleus, which is enclosed by a membrane, and by the division of the cell into many compartments. All multicellular organisms are eucaryotes. There are in addition many single-celled eucaryotes.

eucyte
A eucaryotic* cell.

exocytosis
The expelling of a substance from a cell by uptake into and release from its membrane.

exon
The coding sequence of a eucaryotic* gene* whose transcript* appears in the completed mRNA* and that is expressed* as protein. Exons are interspersed by introns*, which do not code for expressed protein and are removed from the primary mRNA transcript.

exponential law
A mathematical expression in which a quantity is multiplied or divided by the same factor in equal intervals in time. For example, a population grows exponentially when it doubles every *n* years.

expression
When a gene* is expressed, the corresponding protein is produced; when a *genotype** is expressed, the result is a *phenotype**.

F-pilus
F-pili are thread-like structures on the surface of some *E. coli** bacteria. In conjugation*, the cells equipped with F-pili act as donors of DNA.

feedback
The influencing of a process by its end-product. An example of this is autocatalysis*. Feedback is of central importance in many processes of regulation and control.

fermentation
Anaerobic* metabolism*, producing chemical energy in the form of the cellular fuel molecule ATP*.

fluctuation
In molecular physics, a small variation, the occurrence of which cannot be predicted except on a stochastic* (statistical) basis.

fluorescence
A physical property of some substances that can absorb light at a certain wavelength and re-emit it at a longer wavelength; often used as a method of detection of substances that *fluoresce*. Certain fluorescent dyestuffs bind to nucleic acids, giving a method of detection of these as well.

foot-and-mouth disease
A highly infectious sickness of hoofed animals (cattle, goats, pigs, and sheep) caused by the picornavirus*.

fractal
Self-similar structures, that is, structures whose appearance is the same irrespective of the scale on which they are observed, are called fractals.

free oxygen, free hydrogen
Oxygen (O_2) or hydrogen (H_2) in molecular form, that is, not combined with other elements.

gene
A unit in the DNA double strand of the chromosome*. A gene contains the information required to synthesize a protein.

genetic code
The assignment of groups of three nucleotides* (triplets) to the amino* acids that occur in proteins (see Vignette 8).

genetic recombination
The exchange of genes or segments of genes between hereditary material from the male and the female parent, leading to the formation of a new genome*. Genetic recombination is the basis of sexual inheritance.

genetics
The science of the formation of hereditary characteristics and their transmission from one generation to the next.

genome
The collective term for all the genes in a cell.

genotype
The sum of all the genetic information present in an organism.

glass-fibre optics
The transport of light in 'bundles' along glass fibres over long distances.

glycine
One of the twenty common naturally occurring amino* acids; a basic building block of the proteins.

glycoprotein
A polypeptide* that is linked to a sugar-like molecule.

guanine (G)
One of the bases of the nucleic acids; see *base*.

haemoglobin
The protein complex that is responsible for the red colour of blood. Its task is to bind oxygen and to transport it to all the cells of the organism for use in their metabolism.

handedness
See *chirality*.

hepatitis A
An inflammation of the liver, caused by an RNA virus* possessing a single-stranded genome*, with an incubation period of five to sixty days.

HIV virus
Human immunodeficiency virus. A retrovirus* with a single-stranded RNA genome*, that causes the disease of the immune system AIDS*.

homologous, homology
See Sequence homology.

homopolymer
A macromolecule* that is built up of only one kind of monomer*.

horizontal inheritance
The spread of genetic material by recombination. Contrast vertical* inheritance.

hybrid
A new species that results from the association of the fragments of two

different species. In molecular biology it is a double-stranded molecule in which the bases* of an RNA strand are paired with those of a complementary* DNA strand.

hydrogen bond
A relatively weak chemical bond, which occurs between an electronegative atom (such as oxygen or nitrogen) and an electropositive hydrogen atom that itself is bound to a second electronegative atom.

hydrogen cyanide
The well-known poison prussic acid; chemical formula, HCN.

hydrolysis
The splitting of a chemical compound by water; the elements of water are added to the 'free ends' created by each chemical bond that is broken.

hyperbolic growth
Growth in which a quantity increases by the same factor in ever shorter time intervals. For example, the population of the Earth is doubling in shorter and shorter periods.

hypercube
A many-dimensional array of points analogous to a cube in three dimensions (see Vignette 12).

hypercycle
A cyclic coupling pattern relating individual reproduction cycles (see Vignette 14).

immune system
The highly complex defence system of humans and higher animals that responds to foreign molecules (antigens*) by producing antibodies* against them.

immunoglobulin
Globular proteins; antibody* molecules that react specifically with one antigen*.

incubation
(a) In biochemistry, the holding of a sample at a particular temperature, usually in order to allow a culture of bacteria or viruses to grow or a chemical reaction to proceed. (b) In medicine, the period between infection and the appearance of symptoms.

influenza virus
A virus containing single-stranded RNA; the cause of influenza in humans.

information
One must distinguish between information in the 'absolute' and the

'semantic' senses. 'Absolute' information is defined by probability theory. It is the average number of bits* that (for a given distribution of probability of symbol occurrence) is needed to identify a message. The semantics of a message are related to its meaning, and cannot easily be defined mathematically. Both aspects are encountered in the context of genetic information.

information distance or *mutation distance* or *Hamming distance*
The number of differently occupied positions in two related sequences.

information theory
The science of the storage, transmission and processing of information.

insertion
A mutation* in which one or more new nucleotides* are incorporated into a nucleotide chain.

instructed synthesis
Synthesis according to instructions on a template; a term appplied usually to the synthesis of RNA or proteins.

intron
An intervening sequence in a mosaic gene*; in contrast to an exon* it carries no genetic information and its genetic function is at present unknown.

inverted gene fragment
A gene* fragment that is built into its chromosome* with the incorrect direction of reading.

isomers
Chemical compounds with the same number of atoms distributed differently in space or with respect to one another.

iteration
Repetition of a procedure.

kinetics, chemical
The science of the laws governing the rates of chemical reactions.

Krebs cycle
Named after its discoverer (also called the tricarboxylic acid cycle). An enzymically governed reaction cycle in which carbohydrates are oxidized in successive steps to give carbon dioxide. The overall reaction consists of the conversion of one pyruvic acid molecule and three water molecules to three carbon dioxide molecules and ten hydrogen atoms; the latter react in turn with oxygen by further enzymic steps, ultimately giving water. This reaction cycle is the most important producer of energy (ATP*) in aerobic* cells.

laser

Acronym for light amplification by stimulated emission of radiation. A light amplifier. In the laser, individual wavelengths compete and reinforce each other by stimulating emission in phase with themselves. As in selection, a single oscillation mode 'wins', so that the output of the laser is light of a single wavelength and phase. This gives the possibility of attaining extremely high light intensity and sharp focusing.

laser fluorimeter

An apparatus for the measurement of fluorescence* with a laser* as light source.

laser mode

An oscillation of given phase and wavelength in a laser*.

law of mass action

See *mass action, law of.*

laws of thermodynamics

See *thermodynamics, laws of.*

lipid

A collective name for a chemically diverse class of biological compounds, largely insoluble in water. Alongside membrane proteins, lipids are a major constitutent of biological membranes.

lysis

The destruction of a cell by disruption of its cell membrane.

macromolecule

A 'giant molecule'; a polymer*. Important examples in biology are the proteins and the nucleic acids; others include starch, a sugar polymer.

mass action, law of

A thermodynamic* relationship that lays down the quantities of different reaction partners that can coexist with one another in equilibrium (under given conditions, such as temperature and pressure).

master sequence

Dominant sequence in a quasi-species, often showing the highest selection value among the individuals present. It can in many cases be derived from alignment of various homologous* sequences. It is occupied at each point by the monomer* corresponding to the most frequent monomer in the aligned sequences (and is thus the consensus* sequence).

membrane proteins

Proteins that are part of the biological membrane. They impart specific properties to the various biological membranes.

messenger RNA (mRNA)
RNA that serves as the template* for the synthesis of proteins.

metabolism
The production of chemical energy (ATP*) and biologically useful chemicals in the cell, starting from energy sources that can be either chemical (environmental nutrients) or physical (sunlight).

mitochondrion
The mitochondria are organelles found in the cytoplasm* of all aerobic*, eucaryotic* cells. They house the enzymes of the respiratory chain* and are thus the site of ATP* production.

monomer
The repeating molecular sub-unit of a polymer*; the basic unit from which, by the repetition of a particular chemical reaction, the polymer is assembled.

morphogenesis
The formation of an organism or its organs during the process of ontogenesis*.

mosaic gene
A gene that contains alternating introns* and exons*.

mRNA
See *messenger RNA*.

mutagenicity
The ability or tendency to cause mutation*.

mutation
An inheritable change in a chromosome*.

neo-Darwinism
The evolution theory of Darwin expanded by the laws of genetics* and of population* biology in the first half of this century.

neutral mutant
A mutant that, compared with the wild* type, has no selective advantages or disadvantages.

niche formation
A biological niche is a set of environmental conditions within which an otherwise inferior species can flourish without the pressure of competition. In the widest sense, the niche always contains a stabilizing interaction with the environment or with other species. Competitive pressure can be removed, for example, by independent sources of nutrition.

nucleic acid
There are two classes of nucleic acid: ribonucleic acid (RNA) and deoxyribonucleic acid (DNA). Their structures are described in Vignettes 4 and 5.

nucleoside
A base* of nucleic acid condensed* with a molecule of the sugar ribose* or 2'-deoxyribose* (see Vignettes 4 and 5).

nucleoside triphosphate
A nucleoside* condensed* with a triphosphate group (see Vignettes 4 and 5).

nucleotide
A nucleoside* condensed* with a single phosphate group; the monomer* (repeating unit) of the nucleic acids.

nucleotide sequence
The order of the nucleotides* in a piece of DNA or RNA.

ontogenesis
The development of organisms.

operon
A unit of genes that are regulated together as a single unit.

overlapping genes
In a viral genome*, one and the same stretch of nucleic-acid sequence can be used to encode two different proteins, either in the same or in different reading frames. This allows the genes to overlap.

oxidation
(a) In chemistry, donating oxygen or withdrawing electrons. (b) In biology, a term frequently used to mean the production of useful energy by the stepwise oxidation of energy-rich matter.

ozone
Formula O_3; a form of oxygen that occurs in trace quantities in air.

ozone layer
A layer in the stratosphere (8–30 miles high) in which the oxygen present is 85–90% ozone*. It protects organisms on earth from the lethal effects of the Sun's ultraviolet radiation. Its existence was a prerequisite for the development of higher forms of life on our planet.

palindrome
A series of letters that is the same whether it is read from the left or from the right. When this definition is applied to DNA, a different strand is read in each direction.

peptide, polypeptide
A linear molecule formed by the condensation* of of two or more amino* acids with the expulsion of water (see Vignette 6).

phage
See *bacteriophage.*

phagocytosis
The uptake of particulate matter into a cell, in contrast to the uptake of single, dissolved molecules.

phenotype
The result of the expression* of the genes in an organism; the organism itself with all its characteristics. The phenotype is determined both by the genotype* and by the conditions under which the genotype is expressed.

photosynthesis
The synthesis of energy-rich organic molecules from energy-deficient inorganic molecules with the help of the light of the Sun, transformed into an electrochemical potential. The starting materials for photosynthesis in green plants are carbon dioxide (CO_2) and water (H_2O), from which the ultimate products are oxygen (O_2) and glucose ($C_6H_{12}O_6$) or its polymer*, starch.

phylogeny
(a) The evolutionary history of all species. (b) The history of the development of particular interrelated groups of species.

picornavirus
An RNA virus* with a single-stranded genome*.

plasmid
An independent unit of hereditary material that lives in symbiosis with a cell. Some plasmids are replicated together with the cell's nucleus, and others proliferate independently.

point mutation
A mutation* based upon the replacement of a single base-pair* in DNA. This is an alteration at a single point, in contrast to the change caused by a single deletion* or insertion*, which can result in a displacement of the reading frame*.

Poisson distribution
A limiting case of the binomial distribution in which the probability of a particular event is very small but the number of chances that it gets to occur is very great. If we need to know the chances of appearance of a mutant with a given number of errors, then the Poisson distribution is used to calculate this on the basis of the average error frequency.

poliomyelitis
An infection of the central nervous system (usually of the spinal chord) caused by the polio virus, one of the picornaviruses*.

polycistronic
This is the term given to a messenger RNA sequence that contains several genes in a row. In general, genes that are co-regulated in operons* are transcribed* in the form of polycistronic mRNA*; precursors of tRNA* and rRNA* can also be produced initially as polycistronic RNA.

polymer
A macromolecule* built up from monomers*.

polymerase
A group of enzymes* that catalyse* the synthesis of polynucleotides from the energy-rich monomeric nucleoside triphosphates.

polyoma virus
Small DNA viruses that as a rule produce latent and unnoticed infections in their host organisms (humans, monkeys, etc.) and are not oncogenic. Their property of transforming* cells when injected into suitable experimental animals together with the small size of their genomes* makes them important objects for biological research.

polypeptide
See *peptide.*

population biology
A branch of ecology concerned with the influence of the environment upon entire populations of a particular species of plant or animal. Mathematical population biology is the investigation of growth, adaption and the spreading of genes in Mendelian populations by the use of mathematical models.

population number
The number of individuals in a population, usually without consideration of the space it occupies; sometimes, alternatively, the number of individuals per unit of volume (the latter is correctly termed the 'population density').

porphyrin complex
A compound consisting of a metal ion embedded in a ring with the porphyrin structure, a constituent of several enzymes*. These complexes include the cytochromes* and haemoglobin*, in which the metal ion is iron.

prebiotic synthesis
The synthesis of important basic biochemical monomers, such as amino* acids or nucleotides*, under conditions corresponding to those that prevailed on the primitive (prebiotic) Earth. One of the first prebiotic syntheses was carried out by H. Urey and S. Miller.

primary structure
The sequence of covalently linked amino-acid monomers* in a protein (also called the *sequence* of the protein).

procaryote, procaryotic
A single-celled organism identified in particular by the absence of a cell nucleus enclosed by a membrane. The cells are not divided up into compartments and membrane-enclosed organelles are not found.

procyte
A term used to describe the organizational pattern of the procaryotic* cell.

protein
Formerly called albumins on account of their presence in egg-white. The most important functional molecules of the cell. Their basic structure is the polypeptide* chain, which is a polymer* made up of between one hundred and several hundred amino-acid* monomers*. The chain folds up in a characteristic manner, thus bringing different functional groups into close proximity and forming a catalytically active centre* (see Vignette 6).

provirus
The stage in the life-cycle of a retrovirus* in which the virus, integrated into the chromosome* of the host cell, multiplies by replication along with the host cell.

purine
A class of nucleic-acid bases, whose members are adenine and guanine.

pyrimidine
A class of nucleic-acid bases, whose members are cytosine, thymine and uracil.

Q_β virus
An RNA phage* with a single-stranded genome*, whose host is the bacterium *E. coli** (see Vignette 13).

quantum mechanics
A mechanical theory developed by Werner Heisenberg, Pascual Jordan, Paul A.M. Dirac, and others to describe small particles in a manner consistent with itself and with quantum theory. The complementary aspect of this theory is wave mechanics, based upon Louis de Broglie's wave model of material particles, and was developed by Erwin Schrödinger.

quasi-species
The quasi-species represents a weighted distribution of mutants centred around one or several master sequences*. It is the target of selection* in a system of replicating individuals that replicate without co-operating with one another (RNA molecules, viruses, bacteria). In evolution theory, it replaces

the wild* type, which was regarded as the target of selection in the classical interpretation of selection.

quaternary sequence
A sequence containing four classes of symbol.

rabies
An infectious and fatal disease caused by a rhabdovirus*. It attacks principally foxes, rats and domestic animals, whose bite can also infect humans.

racemic
A system is racemic when it consists of a mixture of equal proportions of left- and right-handed derivatives (enantiomers*) of the same compound.

rate law
The way in which the rate of a chemical reaction depends upon the concentrations of the reacting molecules.

reading frame
They way in which symbols are divided up into groups for reading. In the translation* of mRNA*, the sequence ...GGGUUUAAA... can be read in three possible frames: ... GGG UUU AAA ..., ...GG GUU UAA A..., and ...G GGU UUA AA... . In general, only one of these is correct.

recombination
See *genetic recombination.*

reduction
See *oxidation.*

redundancy
In information theory*, the repetition of information already communicated.

reoviruses
A family of animal and plant viruses that possess a double-stranded RNA genome* in 10 to 12 segments.

replicases
The polymerases* that participate in the replication of DNA and RNA.

replication
The doubling of a nucleic acid molecule to give two identical products, normally with the help of an enzyme*.

replicator
General term for any individual (DNA molecule, virus, etc.) that is self-reproducing. A replicator arises in the first place by the mutation* of a replicator already present and multiplies by self-replication only.

repression
The blocking of the reading of a gene* by a repressor*.

repressor
A protein molecule that blocks the reading of a gene* by binding to the corresponding operator.

respiratory chain
The sequence of biochemical reactions associated with respiration.

respiratory enzyme
An enzyme* that plays a role in the respiratory chain*.

restriction enzyme
An enzyme* (endonuclease) that cuts DNA at exactly defined places (see Vignette 15).

retrovirus
A family of RNA viruses* that infect humans and other vertebrates. Their single-stranded RNA genome* is transcribed* into double-stranded DNA (a provirus). This step, which is the opposite of the usual information transfer from DNA to RNA, is catalysed* by a viral enzyme* called RNA-dependent DNA polymerase* or reverse transcriptase*.

reverse transcriptase
An RNA-dependent DNA polymerase* that catalyses* the synthesis of DNA chains with RNA as template*.

revertant
A mutant identical with the wild* type, that has arisen from the mutation* (*reverse mutation*) of a sequence very similar to the wild type.

rhabdovirus
A widely scattered family of RNA viruses* with a rod-like shape. One of these is the virus that causes rabies*.

ribonucleic acid
See RNA.

ribose
A simple sugar molecule containing five carbon atoms (a *pentose*); the sugar component of the nucleotides* from which ribonucleic acid (RNA) is assembled.

ribosome
A cellular particle, containing both protein and RNA, responsible for the translation* of genetic information.

RNA
Ribonucleic acid; differs from DNA in its use of the sugar ribose* (instead of

deoxyribose*) and the base U (instead of T), and also in that it usually occurs with a single-stranded rather than a double-stranded structure. Has roles as a structural element, as a functional entity and as an information carrier. See Vignettes 4 and 5.

RNA phage
A phage* whose genome* is composed of RNA.

RNA polymerase
The polymerase* responsible for the transcription of DNA into RNA.

RNA virus
A virus whose genome* is composed of RNA.

rRNA
The RNA component of the ribosome*.

selection
The fixation of the genotype* by the continual evaluation of the phenotype*.

selection value
A parameter characteristic of a replicating species, mutant or quasi-species, defined by dynamic properties; it takes account of the rate and quality of reproduction and of the lifetime of the individuals. The selection value determines the outcome of selection.

selection pressure
Pressure exerted by the environmental conditions favouring or necessitating adaptation.

sequence
See *primary structure, nucleotide sequence.*

sequence homology
Sequences are said to be homologous when corresponding positions in them are occupied by identical monomers*.

sequence space
A multidimensional hypercube* used for the theoretical representation of all possible variants of a sequence (see Vignette 12).

serial transfer
The systematic transfer of a portion (an *aliquot*) of one sample to the contents of the next, and so on.

sexual recombination
The basis of sexual propagation. The fundamental step in this is the combination of the nuclei from the two parents. An important aspect of the process is genetic recombination* (see Vignette 15).

splicing
The excision of intron-coded* segments from the freshly transcribed* messenger RNA of a mosaic* gene followed by the covalent linking of the exon-coded* segments on each side. This processing is characteristic of eucaryotic* cells.

sporulation
The formation of spores.

stationary state
Term denoting a state in which no changes can be observed with the passage of time.

statistical geometry
A method for calculating degrees of kinship in the analysis of DNA, RNA, and protein sequences. It is based upon a geometrical representation of the distance relationships in sequence space (see Vignettes 1, 2, and 12).

statistical significance
A manifestation of regular behaviour is said to be statistically significant when it can be distinguished clearly from the background 'noise' of random events.

stereoisomers
Isomers* that contain the same chemical groups but that have these in different spatial dispositions.

stereospecific
Specific with respect to the spatial arrangement of the components of a chemical compound.

stereospecificity
The ability of enzymes* to select one out of the two or more stereoisomers* of a substrate, to bind it and to catalyse* its reaction, while ignoring the other(s).

stochastic
Dealing with the random behaviour of individual participants in a dynamic* (time-dependent) process; describable only by statistical methods.

symmetry breakage
Symmetry is the ordered repetition of identical elements of structure. A change in the physical environment can remove the condition that makes symmetry possible. Examples of symmetry breakage are provided by certain phase transitions.

synapse
The junction between nerve threads and nerve cells, across which communication takes place.

T phage
A collective term for seven different virulent phages* of *E. coli**, which on account of their use as model systems for the analysis of replication and genetics have been important in molecular biology.

tautology
A repetition of the same statement or idea in other words.

teleology
The philosophy of final purpose and of purposiveness.

teleonomy
An expression applied first to biology by Monod: structure, performance, or activity that contribute to the success of a biological project are called 'teleonomic'. This term expresses the *seemingly* goal-directed nature of the phenomenon in question.

template
DNA and RNA molecules that contain information in the form of a particular nucleotide* sequence and that are used for the synthesis of further molecules containing this information by a process of complementary* copying.

thermodynamics
The science of heat and work, and the relationship between the transfer of heat and the change of state of the system under study. Thermodynamics also concerns equilibria within material systems ('system' simply means the entire matter under consideration).

thermodynamics, laws of
These are axioms* of physics. The first law of thermodynamics is the principle of conservation of energy. The second, the law of entropy*, states that spontaneous processes in an isolated system are always associated with an increase in entropy. It implies that heat can never be transformed into useful work without some wastage.

threshold value
An exact, critical value; the behaviour of a system can be profoundly influenced according to whether or not the threshold value is exceeded. See also *error threshold*.

thymine (T)
One of the bases of the nucleic acids; see *base*.

tobacco mosaic virus
A plant virus with a single-stranded RNA genome*.

topography
The exact and detailed graphical representation of the physical and geometrical properties (such as height) of a region.

topology

The science describing the properties of sets of points (surfaces, volumes etc.) in two, three, or more dimensions which are retained when the sets of points are mapped.

transcription, transcript

Transcription is the rewriting of the genetic message of DNA into RNA*, resulting in a *transcript* of the information in the DNA.

transfer RNA (tRNA)

Relatively small RNA molecules (about 70–90 nucleotides). Each binds a particular amino* acid and, by means of the interaction between its anticodon* and the genetic message, determines the incorporation of the amino acid at the correct point in the growing protein.

transformation

In genetics, a genetic modification that arises because the cell in question incorporates DNA from another cell or a virus, or acquires a plasmid*, and treats the foreign DNA in subsequent replication cycles as though it were its own.

translation

The translation of the genetic message from messenger RNA* into the corresponding amino-acid sequence of a protein.

transposition

The transfer of a piece of DNA from one place within the genome* to which it belongs to another place in the same genome. Such DNA regions move about relatively frequently and cannot be generally associated with a particular location.

triphosphate

In biochemistry, an abbreviation for 2'-deoxyribonucleotide-5'-triphosphate or for ribonucleoside-5'-triphosphate (see Vignette 3).

tRNA

See *transfer RNA*.

turbidostat

A flow reactor for the cultivation of microorganisms. Fresh nutrient solution is added continually, and at the same speed microorganisms are withdrawn along with the spent nutrient medium. A flow equilibrium (stationary state*) is set up, with a constant population of microorganisms, the population being measured by turbidity. See Vignette 11.

uracil (U)

One of the bases of the nucleic acids; see *base*.

valine
One of the twenty common naturally occurring amino* acids; a basic building block of the proteins.

value function
The selection value* as a function of its physical parameters and variables.

value peak
A peak in the value landscape; a local maximum of selection value*.

value topography
See *topography*, applied to the value landscape.

vegetative reproduction
(a) Asexual reproduction by cell division. (b) The artificial but economically important reproduction of plants in culture, by cuttings and other methods.

vertical inheritance
Inheritance by descent, 'down' a cell* line. Only direct descendents can inherit advantageous genetic material.

vesicular stomatitis
An infectious disease of cattle, caused by a rhabdovirus*.

virion
The complete, infectious virus particle.

virus
A parasitic, infectious particle that exploits the enzyme* machinery and the metabolism of a host cell in order to multiply. It has its own genetic programme in the form of a DNA or RNA molecule, and it is able to invade the host cell (see Vignette 13).

wild type
The best-adapted genotype* (and thus also the phenotype*) of a species living in a natural environment and representing the majority of the individuals of the species. In a quasi-species, the wild type is the centre of the mutant distribution, and is defined by the consensus sequence.

X-ray diffraction
The structures of molecules, including biological molecules, are determined most precisely by exploiting the diffraction of X-rays by crystals, and it is this method that has given us the most information about the structures of proteins and nucleic acids.

zone refinement
A method for the production of pure crystals free of lattice defects. By accurate control of the temperature, a molten zone is made to pass down the length of the crystal, so that this is just melted and then allowed to cool and recrystallize gradually.

Index

The names of authors appearing in the Notes only have not been included in the index.